NELSON
PRODUCT DESIGN
& TECHNOLOGY
WORKBOOK

VCE UNITS 1–4

JILL LIVETT
JACINTA O'LEARY

Nelson Product Design and Technology Workbook VCE Units 1–4
1st Edition
Jill Livett and Jacinta O'Leary

Publishing editor: Deborah Barnes
Project editors: Aynslie Harper and Kathryn Coulehan
Copy editor: Nick Tapp
Text designer: Jennai Lee Fai
Cover designer: Jennai Lee Fai
Cover images: Shutterstock.com/Max Krasnov;
   Shutterstock.com/DmitriyRazinkov;
   Shutterstock.com/Rattanamanee Patpong;
   Shutterstock.com/binbeter
Permissions researcher: Catherine Kerstjens
Production controller: Emma Roberts
Typeset by: Q2A Media

Any URLs contained in this publication were checked for currency during the production process. Note, however, that the publisher cannot vouch for the ongoing currency of URLs.

Acknowledgements
The publisher would like to credit and acknowledge the following sources for photographs in the table of contents: page iii: Shutterstock.com/Rawpixel.com (top), Shutterstock.com/Anna Ok (middle), Shutterstock.com/Rawpixel.com (bottom); page iv: Shutterstock.com/Ogovorka (top), Astrakan Images/Alamy Stock Photo (middle), Shutterstock.com/Comaniciu Dan (bottom)

© 2017 Cengage Learning Australia Pty Limited

Copyright Notice
This Work is copyright. No part of this Work may be reproduced, stored in a retrieval system, or transmitted in any form or by any means without prior written permission of the Publisher. Except as permitted under the *Copyright Act 1968*, for example any fair dealing for the purposes of private study, research, criticism or review, subject to certain limitations. These limitations include: Restricting the copying to a maximum of one chapter or 10% of this book, whichever is greater; providing an appropriate notice and warning with the copies of the Work disseminated; taking all reasonable steps to limit access to these copies to people authorised to receive these copies; ensuring you hold the appropriate Licences issued by the Copyright Agency Limited ("CAL"), supply a remuneration notice to CAL and pay any required fees. For details of CAL licences and remuneration notices please contact CAL at Level 15, 233 Castlereagh Street, Sydney NSW 2000, Tel: (02) 9394 7600, Fax: (02) 9394 7601
Email: info@copyright.com.au
Website: www.copyright.com.au

For product information and technology assistance,
   in Australia call **1300 790 853**;
   in New Zealand call **0800 449 725**

For permission to use material from this text or product, please email
aust.permissions@cengage.com

ISBN 978 0 17 040040 4

Cengage Learning Australia
Level 7, 80 Dorcas Street
South Melbourne, Victoria Australia 3205

Cengage Learning New Zealand
Unit 4B Rosedale Office Park
331 Rosedale Road, Albany, North Shore 0632, NZ

For learning solutions, visit **cengage.com.au**

Printed in China by 1010 Printing International Limited.
7 8 9 10 11 25 24 23 22

CONTENTS

How to use this workbook v

DESIGN FUNDAMENTALS 1

1 The product design process and product design factors 2
2 The design brief, evaluation criteria and research 8
3 Design and development: Drawing types and techniques 24

UNIT 1: SUSTAINABLE PRODUCT REDEVELOPMENT 48

Area of Study 1: Sustainable redevelopment of a product

4 Starting Unit 1 50
5 Product analysis: Investigating and defining 54
6 Design brief and evaluation criteria 62
7 Research, design and development 64
8 Planning for production 66

Area of Study 2: Producing and evaluating a redeveloped product

9 Production and recording progress 68
10 Evaluation of the product 69

UNIT 2: COLLABORATIVE DESIGN 73

Area of Study 1: Designing within a team

11 The product design process and influencing factors 75
12 Working as a team: Investigating and defining 79
13 Research 87
14 Designing within a team: Design and development 95
15 Planning and production 104

Area of Study 2: Producing and evaluating within a team
  16 Production 107
  17 Evaluation 108

UNIT 3: APPLYING THE PRODUCT DESIGN PROCESS 112

Area of Study 1: Designing for end-user/s
  18 Designer's role, the process, factors and market research 114
Area of Study 2: Product development in industry
  19 R&D, technologies, sustainability, planned obsolescence and manufacturing.................127
Area of Study 3: Designing for others
  20 Activities to prepare for the SAT 144

UNIT 4: PRODUCT DEVELOPMENT AND EVALUATION 145

Area of Study 1: Product analysis and comparison
  21 Attributes, values and priorities for comparison 147
Area of Study 2: Product manufacture
  22 Working on your product and documenting progress 160
Area of Study 3: Product evaluation
  23 Using evaluation criteria to judge the product and suggest improvements 162
  24 User instructions and/or care label...163
  25 Exam preparation..166

DESIGN FOLIO TEMPLATE 169

HOW TO USE THIS WORKBOOK

This workbook is designed for students to use and refer to throughout VCE Product Design and Technology Units 1–4. Introductory and general activities relevant to all units can be found in the 'Design fundamentals' section (see details next page). Many activities and explanations in Units 1 and 2 may also be useful in Units 3 and 4, and vice versa. Students can use the workbook for all four units.

The workbook provides structured activities that satisfy the demands of the *VCE Product Design and Technology Study Design 2018–2022* in a sequential flow. For the study's conceptual/theoretical content, students should refer to the *Nelson Product Design and Technology VCE Units 1–4*, 4th edition, textbook by O'Leary and Livett (references to relevant pages are clearly marked) or material from other sources.

Folios for Units 1 to 4

For all units, it is recommended that students use a display folder (A4 or A3) to collect and display their work for assessment. Students can submit the activities and the Design Folio Template sheets as their final work, or use the sheets as drafts or guides and develop more individual or refined work for assessment. Many of the written requirements, particularly tables or journal sheets, can be re-created to suit a student's individual needs, and added to their folio.

The Design Folio Template

Students can use the sheets in the Design Folio Template to help them form a folio for the School-assessed Task (SAT) in Units 3 and 4. The Design Folio Template includes a checklist to ensure that all SAT items for assessment are covered. **This template is a guide**. Students can individualise the work in their folio by developing their own style and format and by emphasising different stages of the product design process relevant to their design situation and adding pages where needed. Certain pages of the Design Folio Template can also be used for folio work in Units 1 and 2. Students should also refer to the current Assessment Criteria for the SAT in Units 3 and 4, published in the *VCAA Bulletin* every year in February.

Exam preparation

The examination preparation section provides areas for revision and suggested revision exercises. The workbook activities can also be used for exam revision.

Solutions to questions

'Correct' answers are not supplied for the activity questions as many are open-ended or relate to the student's individual design situation or the product being analysed – there are many possibilities. It is important that students learn the terminology, think about it and learn to apply it relevantly to specific situations, products and materials. This is particularly applicable to Unit 3, Outcome 2.

'Design fundamentals' section

This section covers much of the core knowledge and skills of the study, and contains activities for introducing concepts in Units 1 and 2. These can be reused in Units 3 and 4, particularly those relating to:
- the product design process
- material classification and materials testing
- the product design factors
- research
- product analysis and evaluation
- sustainability
- design brief and evaluation criteria
- drawing techniques and exercises
- design elements and principles
- risk assessment
- critical and creative design thinking
- log or journal formats.

Order of activities

Activities are presented in a sequential order (check the calendar at the start of each unit) to cover the requirements of the Study Design; however, teachers can change the order or choose which activities are most suited to their students.

Important features of the workbook

DFT Refers to pages contained in the Design Folio Template. Some of these pages are suitable for Units 1 and 2. All pages are used to form the SAT folio in Units 3 and 4.

A4 Indicates this should take an A4 page in your folio.

A3 It is recommended that these pages are photocopied and enlarged to A3 size (141%).

bespoke
Terms that appear in bold brown type are those that are defined in the Glossary on the student book website.

DESIGN FUNDAMENTALS

This section will cover the following areas in detail:
1. The product design process and product design factors   p. 2
2. The design brief, evaluation criteria and research   p. 8
3. Design and development: Drawing types and techniques   p. 24

CHAPTER 1

The product design process and product design factors

Unit 1, Outcome 1, also suitable for: Outcome 1 of both Units 2 and 3
1.1 The product design process..p. 2
1.2 Product design factors...p. 3

The product design process and the **product design factors** are cross-study specifications that are learnt and used in all four units of VCE Product Design and Technology. The activities in Chapter 1 can be used as an introduction and repeatedly as refreshers for all units.

1.1 THE PRODUCT DESIGN PROCESS

No two designers work in exactly the same way or carry out design activities in the same order, but there are recognisable and common stages that they adapt and use as their situation requires.

For this study, the product design process has been defined as having four stages, with each stage having a number of steps.

Activity 1.1a: The stages in the product design process

Refer to the product design process diagram on page 10 of the *Product Design and Technology VCE Study Design 2018–2022*. (For information on the product design process, see pages 8–51 of your Nelson textbook.)
1. Choose a colour to represent each of the four stages of the product design process. On the diagram below, lightly colour the correct number of steps of pie slices.
2. Write the name of each stage *outside* the correct segment (or indicate with an annotation/arrow).
3. Number and write the steps (activities for a designer) *inside* or near each pie slice in the order given in the Study Design.

> **TIP**
> Stage 1 = 4 steps
> Stage 2 = 3 steps
> Stage 3 = 2 steps
> Stage 4 = 1 step

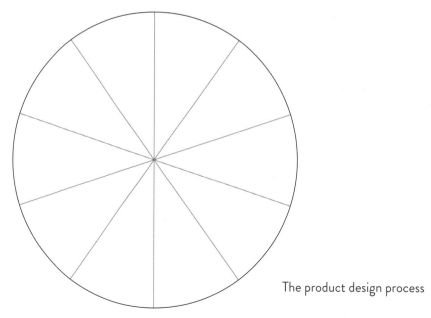

The product design process

Activity 1.1b: The product design process

1. What is the initial step of the product design process?

   _____

2. Give some suggestions as to why a designer (or yourself as student designer) would carry out research such as tests or trials on **materials** (for availability, pricing, source and sustainability), at the following stages:
   - when identifying the need

   _____

   - during the design and development stage

   _____

   - when halfway through production.

   _____

3. Suggest why a designer might draw more design ideas during production.

   _____

4. Explain why the product design process is seen as a circular process with steps that may change in order (not a definite linear process in which each step must follow the other).

   _____
   _____
   _____

5. **Evaluation** involves looking at an object, an activity or a situation and making judgements about what is positive, negative, successful or not successful, what should go ahead and what should change. During which steps could this happen, other than the final evaluation?

   _____
   _____
   _____

1.2 PRODUCT DESIGN FACTORS (FOR ALL UNITS)

Activity 1.2a: The product design factors

1. During which steps of the product design process is it crucial to consider the product design factors?

   _____
   _____

2. Explain the meaning of the word 'parameter' in the context of the product design factors (see page 11 of the *VCE Product Design and Technology Study Design 2018–2022*)

   _____
   _____

3 On the following table, briefly summarise the parameters in the second column.
In the third column, write two or more phrases explaining what a designer would need to research, explore or consider for five or more product design factors if they were **designing a suitable school bag for your school**.

Product design factors	Parameters (summary)	When designing a schoolbag
Purpose, function and context		
User-centred design		
Innovation and creativity		
Visual, tactile and aesthetic (design elements and principles)		
Sustainability		
Economics – time and cost		
Legal responsibilities		
Materials: characteristics and properties		
Technologies: tools, processes, and manufacturing methods		

(Refer to the list of product design factors on page 91 and the explanations of them in Chapter 4 of your Nelson textbook. You will also find a list of product design factors and their 'parameters' on page 11 of the *VCE Product Design and Technology Study Design 2018–2022*.)

Activity 1.2b: Sustainability

Read the explanation of sustainability on page 119 and perview the rest of Chapter 5 of your Nelson textbook.

1 What does 'sustainability' mean?
_____
_____

2 The three main aspects, or 'pillars', of sustainability are listed below. Explain what each of these means when discussing sustainability.
- Environmental _____
_____

- Social _____
_____

- Economic _____
_____

3 Draw a Venn diagram in the space below to show the interconnectedness of the three areas, and create a symbol or image to explain each area.

4 Life cycle assessment assesses a product's impact on the environment at all stages, from raw materials to its end of life. (Remember that this can also be called life cycle analysis.) Research LCA online or read pages 123–4 in your Nelson textbook. In the table below, describe at least two possible areas for sustainability improvement for each stage of a product's life (e.g. a schoolbag or the product you will make).

Product: _____

Stage of a product's life for LCA	Consideration for improved sustainability
Materials sourcing	

Stage of a product's life for LCA	Consideration for improved sustainability
Manufacture	
Transportation	
Product use	
End of product's life (disposal)	

Activity 1.2c: Sustainability questions

1 Why does a durable (long-lasting), high-**quality** product have less impact on the environment than one that is disposed of more quickly?

2 Why would looking after a product carefully so it lasts longer be considered a sustainable practice? How could a manufacturer help a user to do this?

3 What are some of the problems connected with landfill?

4 Identify two materials that you will be working with and explain how they could be disposed of, e.g. sent to landfill, recycled into an inferior material or reused in another product.

5 Single-use plastics are considered a big sustainability issue. List five or more products you can think of that are 'single-use', e.g. drinking straws.

_____

_____

6 Choose two of the products you identified in Question 5, discuss with your classmates and suggest ways in which these products could be: used more effectively before being discarded; replaced with a different product that can be used many times; eliminated with different behaviour.

_____

_____

_____

_____

_____

_____

_____

(Refer to Chapter 5 of your Nelson textbook, on sustainability.)

CHAPTER 2

The design brief, evaluation criteria and research

Unit 1, Outcome 1

2.1	Exploring details for the design brief	p. 8
2.2	Developing evaluation criteria	p. 9
2.3	Planning your research	p. 10
2.4	Research for inspiration	p. 11
2.5	What materials could you use?	p. 13
2.6	Tests and trials	p. 17
2.7	Research into sustainability	p. 23

2.1 EXPLORING DETAILS FOR THE DESIGN BRIEF

Before writing your design brief, it is helpful to explore the 'who, what, when, where, …' of the design situation. This information can be used to write your design brief – there are specific instructions explaining how to do this in each of the units.

Activity 2.1: Design brief web

Fill out the design web on the following page to sort out the details of your design situation. The examples below may help you.

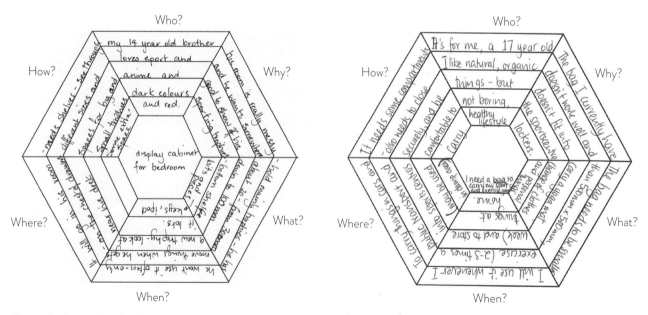

Example design brief webs

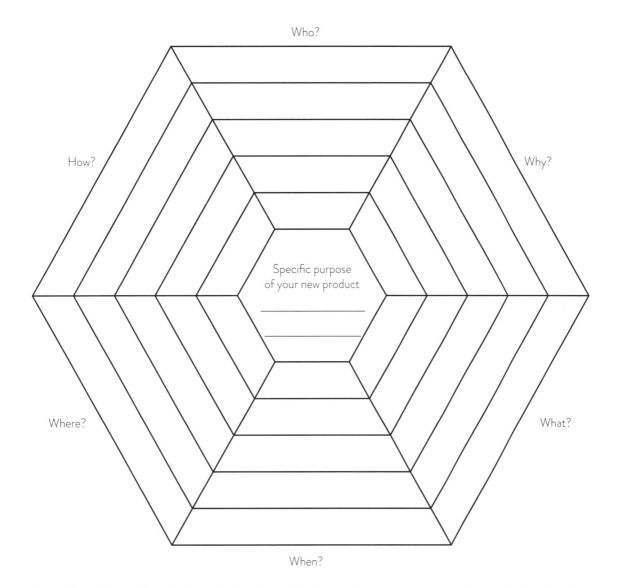

Who will use the product (individual, family), and what are their ages, interests, the styles they like?
Why do they need the product?
What does the product need to do?
When and how often might you use the product?
Where might the product be used, and under what conditions?
How might it work, and what new features could it have?

2.2 DEVELOPING EVALUATION CRITERIA

Evaluation criteria come directly from the design brief and are directed at both the design option drawings (to select the preferred option) and the finished product.

In Unit 1 be sure to cover aspects of the product's redevelopment and the expected improved sustainability. See pages 84–6 for Unit 2, pages 174, 176–7 for Unit 3 and in all units include the due date and budget.

> **TIP**
>
> When writing about evaluation criteria, it is important to get the endings correct: several criteria (plural), one criterion (singular).

(For information on evaluation criteria, see pages 19–21 of your Nelson textbook.)
For the design options, the evaluation criteria are needed as questions only.

Evaluation criteria examples for the design option drawings

- Does this design option incorporate the functions for …?
- Will this design option fit the …?
- Do the shape, style and colour in this option suit the user?
- Are the suggested materials in this option suitable?
- Are the construction methods suggested in this option of a suitable quality?
- Will the changes suggested to improve the existing product's sustainability in this option be suitable for the product?
- Can this option be made in six weeks?

For the finished product, they are written in four parts:
1. a question directed at the finished product
2. a justification (reason why it is important)
3. how to achieve this (suggest research areas, design activities)
4. a description of how you will check this once the product is finished.

Evaluation criteria examples for the finished product

- Does the product function in the way the user needs it to?
- Does the product fit the …?
- Do the shape, style and colour suit the user?
- Is the cost of the product within the budget limit of $____?
- Is the finish of the product of very high quality? Were the joins constructed accurately and are they strong?
- Were the areas I redeveloped (list them) a significant improvement? Is my product better than the original product design, particularly in terms of its sustainability?
- Did I follow my timeline and finish my product in the time allowed?
- Was the cost of my finished product within my budget?
- Are the properties and characteristics of the materials suitable?
- Did I finish my product with a high level of care and attention to detail?

Activity 2.2 Writing your evaluation criteria

Write a series of evaluation criteria for your designs and finished product based on the information in your design brief. Use the product design factors as a prompt to make sure you have considered the areas that need to be covered. Use the evaluation criteria template on page 177 in the Design Folio Template to help structure your evaluation criteria format and content.

2.3 PLANNING YOUR RESEARCH

Research for inspiration and information can be done in many ways. It can also be required at many stages during the design process.

Activity 2.3a: Thinking about research

Tick as many of the following research activities as are appropriate for your design situation.

Secondary sources (desk work!)

- ☐ Look at the work of other designers.
- ☐ Collect printed or digital images of objects, shapes, textures, etc. of interest.
- ☐ Read and apply the research carried out by experts on aspects such as the source of materials, **characteristics and properties**, etc.
- ☐ Search out suppliers and costs for equipment, components and parts and compare what they offer.
- ☐ Compare the costs of different materials.
- ☐ Refer to subject-specific texts on how to complete processes or construction techniques.

Primary sources (your personal research!):

- ☐ Take photos of (or draw) natural or constructed environments or objects.
- ☐ Take required measurements.
- ☐ Try out different processes (practical trials) as a way of improving your skills.
- ☐ Test and explore different materials to become familiar with their characteristics and properties.
- ☐ Think about what materials will look like and how well they 'function' in the product.
- ☐ Other
- ☐ Observe and talk to people using similar types of products.

Activity 2.3b: Planning research and design activities

The research that you carry out should be relevant; targeted to the needs of your situation. Use page 178 of the Design Folio Template or create your own graphic organiser to:
- suggest categories for researching and exploring (refer to your design brief requirements) and list specific areas within those categories. (Categories could include materials, sustainability, how similar things function or are constructed, functional aspects to include, things that will need to be measured, designs to explore, ergonomic information, possible processes and skills needed, etc.)
- highlight the areas that are most important for your design work – areas that you will research. (Materials need to be included.)

2.4 RESEARCH FOR INSPIRATION

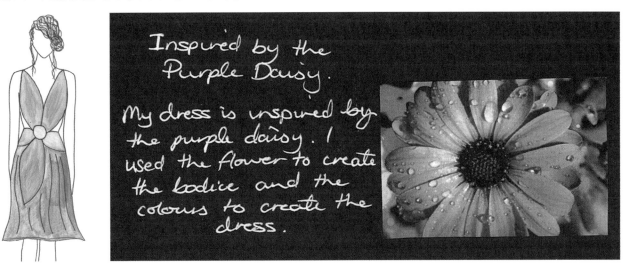

Drawing by student Rachel Krause inspired by a flower

Activity 2.4: Inspirational images of existing products or objects

Find four images of products or objects, two related to your design focus and two from another field (on the Internet, in books, catalogues, magazines, or from your own photos). Set out an A4 or A3 page as shown in the following table, and attach or insert the images.

Annotate each image (make comments) to explain:
- the use of design elements and principles (line, shape, colour, texture, balance, proportion, symmetry), e.g. the thick yellow line around the edge contrasts with the bulk of the black body)
- the use of materials – suitability, contrasting, etc. (name the materials)
- the construction techniques used and any components
- how the product might function – what its main purpose is
- where it might be appropriate for (e.g. outside, inside, formal or casual setting/environment)
- quality and finish
- what you could use/adapt/modify as part of your design (or redevelopment in Unit 1).

Acknowledge the IP of the creator (i.e. name of designer, brand or manufacturer, store) and include the source of your image, e.g. URL, book, magazine, etc. See pages 52–3 in Chapter 4 for more on IP.

A3

Research: Inspirational images and some visualisations		
Image 1: Created by:	My own ideas (visualisations – see page 64):	Image 2: Created by:
Image 3: Created by:		Image 4: Created by:

Change the layout to suit your presentation, allowing enough room for your own visualisation drawings.

2.5 WHAT MATERIALS COULD YOU USE?

Research on materials may be needed before any designing begins, after the preferred design option is selected or even at certain stages in production. In this workbook, the research activities are given before designing, but they can be undertaken when it suits your situation.

(You will find specific information about materials classification, characteristics and testing in Chapters 6, 7 and 8 of your Nelson textbook.)

Activity 2.5a: Material categories and classification

Research is required for this activity

1. Complete the following table by explaining the reasons for each sub-grouping and placing the materials from the lists into their correct sub-groupings. Some reasons have been completed for you.

Wood, metal and plastics categories

Category	Sub-grouping	Reasons	Materials
Wood	Softwood	Comes from softwood trees, needle-like leaves, cones, simple cell structure	
	Hardwood		
	Processed timber		
Metal	Ferrous		
	Non-ferrous	Metal that does not contain iron	
	Alloys		
Plastics	Thermoplastic	Becomes soft and easily moulded when heated, sets in shape when cooled	
	Thermoset		
Composites			

Wood, metal and plastics materials list

acrylic sheet
aluminium
Bakelite
blackwood
brass
bronze

cedar (not Australian)
copper
ebony
epoxy resin
fibreglass
kauri

melamine
mild steel
particle board
plywood
polyester
polystyrene

PVC
radiata pine
silver
titanium
Victorian ash (eucalypt)
zinc

Fabric, fibres and yarns categories

Category	Sub-grouping	Reasons	Materials
Natural	Protein		
	Cellulose		
Manufactured	Regenerated	e.g. fibres from a natural source (usually cellulose) that have been chemically altered	
	Synthetic		
Blended fibres			
Laminated (layered) or treated			

Fabric, fibres and yarns materials list

cotton	lycra	polyester cotton	Gore-Tex®
flax	nylon	PVC	Tencel®
hemp	polyester	rayon	wool

Activity 2.5b: Types of fabric construction

Characteristics and properties of a fabric are not only defined by the fibre used (fibre content) but are also influenced by the way the fabric is constructed. It is important to know how a fabric is constructed to analyse whether it will be appropriate for the needs of the situation.

Write the fabrics listed underneath the following table in the correct column. Attach real examples of each where possible.

	Woven	Knitted	Felted or bonded
Fabrics			

brocade	gabardine	jersey	tricot
corduroy	interfacing	organza	tulle
crepe	interlock	ribbing	twill
felt	jacquard	sateen	

Many of the fabrics listed above can be made from a variety of different fibres or fibre blends. Note that the names given in this list only tell us the construction method, not the fibre being used. When naming a fabric at VCE level, you are expected to give the full name, which includes the fibre content (e.g. 100% wool, 100% modal, 50% acrylic and 50% mohair, etc.).

Activity 2.5c: Fabric construction samples

There are three main types of fabric construction. Create a table as follows to a suitable size, and under each fabric type draw a simple diagram showing two examples of how threads/yarns can be constructed. Attach a sample of each type of fabric construction.

	Woven		Knitted		Felted or bonded	
Name						
Diagram						
Sample						

Material choice

Many factors influence your choice of material. They include the material's appearance, characteristics and properties, cost, availability, useability, etc. Some of this research will be from secondary sources (desk work) and some will be from primary sources (your personal research, e.g. tests and trials).

(For a discussion of factors related to material choice, see Chapters 6, 7 and 8 of your Nelson textbook.)

Activity 2.5d: Materials research

1   List and explain four material characteristics and/or properties that are important for your product. (Some wood, metal and plastic examples are: weatherproof, a particular colour range, level of strength, hardness, ease of working, etc. Some textile examples are: absorbency, drape, elasticity, crease free, weight, etc.)

   a   _____

   b   _____

   c   _____

   d   _____

2   Choose four different materials that could be used for your product. Use the layout shown on the next page to record your research for two samples on one A4 page. Research the following areas:
   - the name of the material, including its raw form where possible
   - how the material is classified
   - a description of its appearance (colour, texture, sheen, patterning, grain, etc.) and at least three properties/characteristics
   - cost (e.g. per metre of relevant thickness or width).

3   Include a sample of each material if it is fabric; and ideally a real sample of resistant material (wood, metal or plastic) if it is thin; otherwise insert a scanned colour image obtained from the Internet.

Materials research

A4

Insert sample (actual or photo)	Insert sample (actual or photo)
Material name: Classification: Cost: _____ per _____ Appearance and properties/characteristics:	Material name: Classification: Cost: _____ per _____ Appearance and properties/characteristics:

Use one A4 page for two samples, or as required.

2.6 TESTS AND TRIALS

A simple materials test will tell you which material performs better in a particular situation.

Choose to test a property/characteristic that is relevant to your design situation. Your choice should be influenced by your design needs and the range of materials available.

(For information on material properties and characteristics, and tests you could carry out, see Chapter 6 of your Nelson textbook.)

Good planning is essential. You need to make sure that the set-up and procedure you use in your materials test will be consistent and give reliable results.

Activity 2.6a: Materials testing, planning and report

Before the test

1. Explain the characteristic/property that you will be testing.

   _____

2. Why is this important for your design situation or your type of product?

   _____
   _____

3. You need to test at least 2–4 different materials. List your materials in the following table.

4. Based on your existing knowledge of these materials, what do you think will happen?

   _____
   _____
   _____

5. Describe, step by step, how you will carry out your test.

   - _____
   - _____
   - _____
   - _____
   - _____
   - _____
   - _____
   - _____

6 Draw a diagram or take a photo of your testing set-up. Label it well, identifying each piece of equipment and explaining how it will work. (You may need to borrow testing equipment, weights, etc. from your school's science department.)

7 Create a table to record your test results. What are you measuring? Is there a numerical value? Have you left space for recording your observations of each material? If the table below doesn't suit your test needs, develop one that does.

Materials	Record of results			Comments and observations

Adjust columns and add rows as required.

During the test

8 Record the results of the test in your table, including your general observations of what happened when each material was tested. You can also record your results on a graph like the ones below, either drawn by hand or by the use of charts in Microsoft Word. This communicates clearly and makes your information more visually engaging.

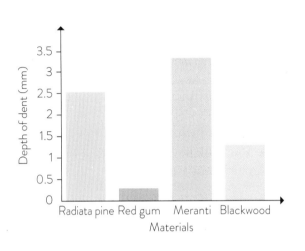

Hardness test results in a bar graph

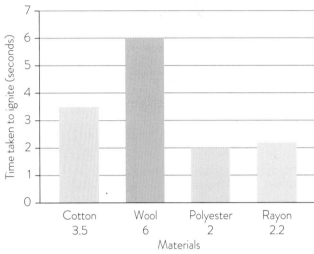

Flammability test results in a bar graph

After the test: Your conclusions

9 What do the results tell you about the materials tested? Which material would be best for your design situation? Why? Which material would be the least suitable? Why?

_____
_____
_____
_____
_____
_____
_____
_____

10 Are the test results what you expected? Why, or why not?

_____
_____
_____
_____

11 Evaluate your testing procedure. Was your testing process accurate and consistent? How could your test have been improved? Could you have used equipment that is more accurate?

_____
_____
_____
_____
_____

Activity 2.6b: Process trials

Research and trial four different processes to make your product (e.g. marking out, cutting/shaping/forming, joining/assembling/constructing product parts, any decorating or embellishing and finishing processes).

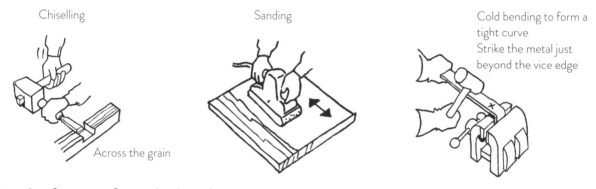

Examples of processes for wood and metal

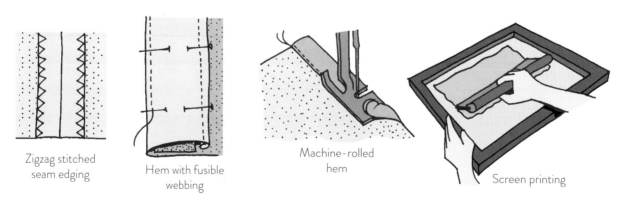

Examples of processes for textiles

Your research will help you decide on the most suitable processes. Record your findings as shown in the table on the following page, or as in Activity 13.4a in Unit 2 on page 92.

Processes research

This table records your secondary research on joins. Adjust columns and add a row as required for each process.

Process 1 – diagram Where or when used? Source of information:	Step-by-step instructions (include quality measures and checks): Advantages/disadvantages:
Process 2 – diagram Where or when used? Source of information:	Step-by-step instructions (include quality measures and checks): Advantages/disadvantages:
Process 3 – diagram Where or when used? Source of information:	Step-by-step instructions (include quality measures and checks): Advantages/disadvantages:
Process 4 – diagram Where or when used? Source of information:	Step-by-step instructions (include quality measures and checks): Advantages/disadvantages:

Evidence of process trials

This table records **your primary research**, trying out and practising new joins and processes. Adjust columns and add a row as required for each process.

Process 1 – photo or sample of trials/attempts	Problems with first attempt:
	Improved by:
Process 2 – photo or sample of trials/attempts	Problems with first attempt:
	Improved by:
Process 3 – photo or sample of trials/attempts	Problems with first attempt:
	Improved by:
Process 4 – photo or sample of trials/attempts	Problems with first attempt:
	Improved by:

2.7 RESEARCH INTO SUSTAINABILITY

Activity 2.7: Materials and sustainability

When comparing the sustainability of different materials, you need to complete research into different aspects of the materials, such as how they are made.

Wood, metal and plastics information

Search for the following terms to find more information online:
- Wood Solutions
- Greenpeace good wood guide
- Forest Stewardship Council Australia
- Plastics Industry (types of plastics)
- American Chemistry Council 'plastics'
- The Australian Steel Industry (Sustainability)
- The Australian Aluminium Council (Sustainability).

Textiles information

- Do an Internet search on 'name of the fabric + sustainability' or 'name of the fabric + embodied carbon'; 'traditional vs organic cotton'; 'microplastics from clothing'; 'fast fashion + environmental impact'; 'LCA wool', etc.

All materials

- For more unusual and recently developed materials, go to the Materia website.
- Go to the DATTA Vic website's sustainability resources and download 'What is eco-design?' – there are Quick Guides for Fashion and for Product Design.
- Go to Business Victoria and on their Sustainable Business page, search for either 'designing sustainable products' or 'designing sustainable fashion'.

Presenting your research

Refer to the two materials that you have tested. You may not be able to find out all the information required below for both, but do your best.

Rate each material with a High, Medium or Low score according to the level of environmental impact or use these symbols to rate: 😊 😐 ☹

Questions about the materials	Material 1 Name:	Rating	Material 2 Name:	Rating
Where it is sourced? • country • method				
How it is processed? • use of energy, water, chemicals • toxic waste • embodied carbon • land degradation				
What distances are travelled? • raw materials to processing • processing to production/retail				
What environmental impacts occur through the use of a product made of this material? • washing • waste created				
Durability – how long will the material last? What contributes to its durability (or lack of durability)?				
Disposal – can it be recycled or re-used? Length of time to break down if biodegradable				
Summing up • Which material appears more sustainable?				

CHAPTER 3

Design and development: Drawing types and techniques

Unit 1, Outcome 1
3.1 Types of drawings .... p. 24
3.2 Visualisations .... p. 25
3.3 Design options as presentation drawings .... p. 28
3.4 Thinking and designing creatively .... p. 30
3.5 Drawing techniques for resistant materials .... p. 36
3.6 Drawing techniques for non-resistant materials .... p. 40
3.7 Working drawings .... p. 43

3.1 TYPES OF DRAWINGS

During the design and development stage of the product design process, there are three steps (5, 6 and 7) with different types of drawings. See the table below. This section on design and development is relevant to all units and explains the different types of drawings as well as some creative and critical thinking techniques.

		Step 5 Visualisations	Step 6 Design options (presentation drawings)	Step 7 Working drawings
Drawings suitable for resistant materials (wood, metal and plastics)				
3D	• 2-pt perspective • isometric • oblique (cavalier or cabinet)	✓ small, quick, of parts or whole with some colour	✓ more detailed, coloured, annotated and whole of product	~ may be included to aid clarity
2D	Orthogonal	✗	✗	✓ scaled, accurate and dimensioned, indicating materials and components
Drawings suitable for non-resistant materials (yarns, fibre and fabric)				
3D	• Using a croquis figure • Fashion illustrations (rendered to show depth or form)	✓ small, quick, of parts or whole with some colour	✓ more detailed, in proportion, coloured, annotated and whole of product	~ may be included to aid clarity
2D	'Flats' or trade sketch	✗	✗	✓ scaled, accurate and dimensioned, indicating materials and components

3.2 VISUALISATIONS

Visualisations are sketches, drawings and/or models to explore ideas. The sketches don't have to be polished – get your ideas down quickly while they are fresh.

(For explanations on visualisations, see pages 29–30, 59–61 and 69 of your Nelson textbook.)

Visualisations by student Demi Spyropoulos with end-user feedback on brown paper

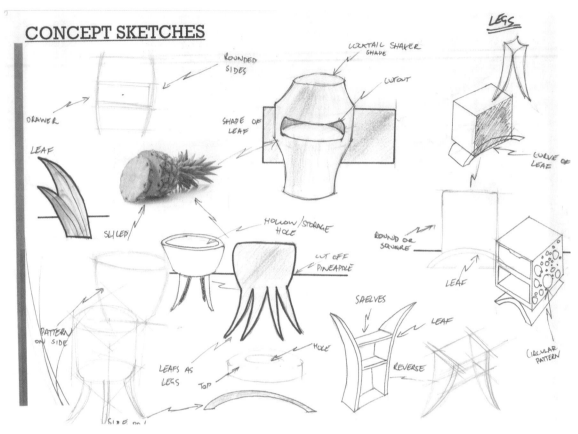
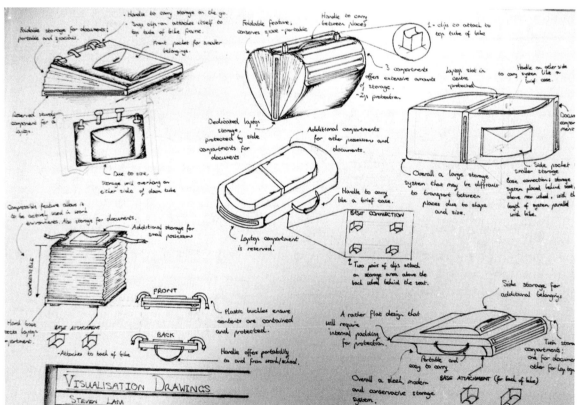

Visualisations based on a pineapple by student Josh Goudge (top) and for a pannier by student Steven Lam (bottom)

Visualisations, inspired by flame, from student Lian Wilson

Critical thinking

Once you have several pages of visualisations, including photos of 3D models, you can apply some critical thinking techniques to decide what to develop further into design options. You could:
- refer to your evaluation criteria, look at and annotate your visualisations and comment on ideas that may not be feasible
- check some facts about what is possible, available and affordable
- run some material tests to check whether ideas will work
- ask people to give you constructive feedback
- decide which ideas look the most aesthetically pleasing or exciting.

3.3 DESIGN OPTIONS AS PRESENTATION DRAWINGS

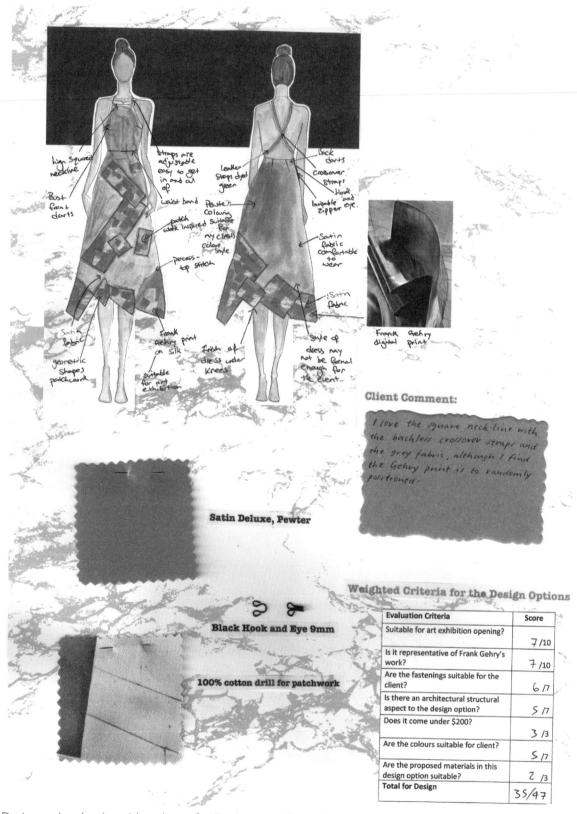

Design option drawing with end-user feedback, material samples and inspiration image by Demi Spyropoulos

DESIGN FUNDAMENTALS

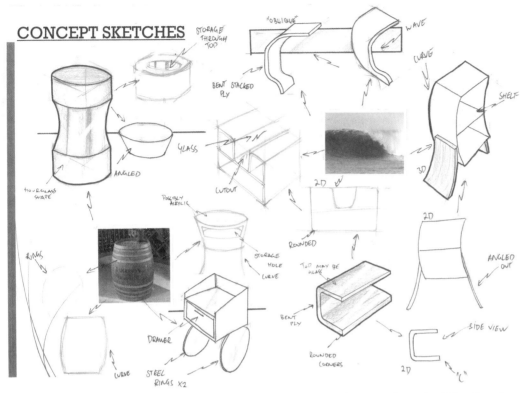

Visualisation drawings, inspired by a wooden barrel and the shape of a wave, from the folio of student Josh Goudge

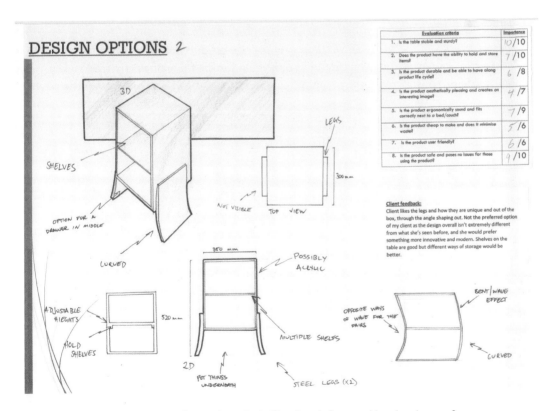

Design option for a storage unit by student Josh Goudge, influenced by the shape of waves

Design options are drawn as 'presentation drawings'. They need to be 3D, clear and fully worked-out designs to show what you intend the product to look like. Use your visualisations as the basis for developing your design options. Refer to the table on page 24 to see which drawings are suitable.

For resistant materials the drawings should either be perspective drawings, isometric or oblique, as shown on the following pages. They should be fairly accurate, so be sure to use drawing equipment (rulers, squares, curves, a 'math-o-mat', etc.) if you don't feel confident drawing them freehand.

For non-resistant materials, they should be illustrations.

Include **annotations** to highlight or describe:
- materials being used
- processes such as type of joins, any fasteners, construction methods
- functional or aesthetic/decorative features
- any fittings or components needed
- **modifications** made to the original design (Unit 1 only)
- how the product fulfils the requirements of improved sustainability (Unit 1), your design brief and evaluation criteria.

3.4 THINKING AND DESIGNING CREATIVELY

Thinking creatively and using ideas in unexpected combinations helps designers to solve problems in unique and innovative ways. The following activities will help you become familiar with design elements and principles, show you how to be inspired and how to expand your design ideas. They can be used when creating your visualisations (as quick, small drawings) or presentation drawings (as more careful drawings showing more detail).
(For more detail on creative and innovative thinking, see pages 102–3 of your Nelson textbook.)

Activity 3.4a: Using design elements and principles

The activities that follow are designed to be quick, around 5–8 minutes each. Check with your teacher which ones you need to complete.

Choose one of the product types in the table below. (You can use one product for all the activities or change products as you go.) For Unit 1, use the product you intend to redevelop for the following activities.

If there is not enough space provided, create your own pages to add to your folio, or photocopy the pages at 141% (A3 size).

Resistant material products		
Jewellery box	Bedside table	Teapot
Child's chair	Kitchen implement	Display case
Skateboard	Small storage shelf or cabinet	Lamp
Desk organiser	Pair of sunglasses	Bookshelf
Garden ornament or furniture	Piece of jewellery	Footstool
Non-resistant material products		
Hat	Top or shirt	Bag
Sleepwear	Jacket (knitwear)	Soft toy
Fitted or loose skirt	Trousers, pants or jeans	Bedding (quilt/sheets)
Dress or pinafore/tunic	Swimwear	Costume

Shapes

A3

1. Choose two shapes – geometric or organic. Use them together in different ways to create three different designs (for the product type you have chosen from the list above). You can repeat the shapes in different sizes in the one frame, e.g. large, medium and small sized ellipses with two triangles.

Line

Lines can be formed by the edges of a material, where two materials or parts of products meet, by seam lines, and by parts such as handles, pockets, etc. They are thick, thin, varied, curved, zig-zag, etc.

2. Find two images of products where lines play an important part or are visually interesting in the design. In the boxes below, draw and annotate the lines in the images to identify and describe them.

Combining line and shape

3 a Using the product type you drew in Question 1, develop three more designs in the boxes below: one that uses geometric shapes and curved/organic lines; one that uses organic shapes and angular lines; and one that uses a combination of shape and line types.

b Which design do you find most interesting and satisfactory? Why?

_____

_____

Colour

4 Choose one of the designs you drew for Question 1 or Question 3a. Draw the design three times (once in each box below) and colour each drawing using a different colour system or combination.

Colour systems or combinations

5 Draw linking lines to connect the colour systems/combinations to the description that they best match.

Colour system/combination	Descriptions
Primary	Close to each other on the colour wheel (e.g. green/blue)
Contrasting/complementary	Browns, cream, beige, earthy reds
Analogous, harmonious	Red, yellow, blue
Neutral	Large proportions of two or more flat, bright and contrasting colours
Classic	Occurring opposite each other on the colour wheel (e.g. purple/yellow) or that have strong tonal differences
Natural	Pale colours (e.g. pink, lemon, mint)
Pastels/gelati	White, black, beige, brown
Metallic	Black, royal blue, bottle green, red, white
Colour blocking	Based on the colours of metals (e.g. silver, gold, bronze, copper)

DESIGN FUNDAMENTALS 33

Using design principles

6 In the top left square of the following table, draw a quick sketch of your original design (chosen in Activity 3.4a Q1 or your product that will be redeveloped in Unit 1), then alter it according to the instructions in each box below.

Sketch the original design	Take an element (line, shape, colour, texture, form etc.) and use it repeatedly
Draw it asymmetrically	Use strong lines or patterns to change the visual impact of the product (e.g. to emphasise an edge or join)
Change some elements of the form so that they become a larger proportion of the whole	Add negative spaces or change the use of negative space

Other changes you could make would be to:
- change the surface qualities (i.e. texture, sheen or pattern) of the material to make it either more complex or more simple
- change some rectilinear (straight) shapes to curved shapes (or the other way around) to change the feeling of 'movement' or rhythm in the design
- use strong contrast in two of the following areas: line, shape, texture, tone, pattern, colour.

Activity 3.4b: Inspiration from elsewhere

One method of thinking creatively is to use objects or things from completely different parts of life, such as animals, plants and architecture, to provide ideas. See the student work on previous pages.

Natural inspiration

Select and use images of nature to inspire the shape, line, texture or colour of your product design.

Garment design, inspired by a Venus flytrap, by student Sarah Luisi

Incorporating architectural elements

Choose a building with strong design features and identify the dominant shapes, lines and colours by drawing over the picture with coloured pencils, textas or highlighters. Use those features in the structure and/or decoration of your product design.

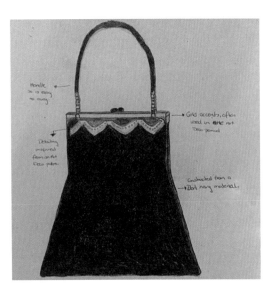
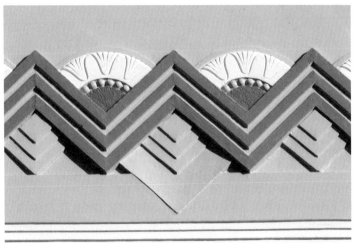

Drawing by student Alex Mocevic, inspired by Art Deco architecture

Reflecting an animal

Many designers use animals as inspiration. Extend your product ideas by reflecting a different animal. Consider using: a simplified shape of the animal; an easily recognised part; the creature's texture and patterning, the line of a part of its body, etc.

Fish shoe by Andre Perugia, 1931, inspired by Georges Braque's fish paintings

Children's furniture animal, inspired by Elad Ozeri, 2010

Drawings, inspired by Frank Gehry architecture, from the folio of student Demi Spyropoulos

Draw up a table as below, insert your inspirational image (animal, plant, building, etc.) in one column and draw your product ideas in the other. Add a row for each inspiring image.

Inspiration: _____	Product: _____

Activity 3.4c: Trigger words

Use trigger words to help you be creative with your designs. Create your own table using the headings below, and make space for the drawings in each column or by adding another row. These triggers can extend your ideas in any of the previous activities with design elements and principles or the inspiration used from elsewhere.

Original design (or one of your ideas)	Subtract – take something significant away	Parody – make fun of the design, turn it into something silly	Combine – merge the idea with another design sketch you have developed

Use some of the other trigger words listed on pages 55–6 of your Nelson textbook.

Trigger word: _____	Trigger word: _____	Trigger word: _____	Trigger word: _____

Create drawings using each trigger word to stretch your creativity.

DRAWING TECHNIQUES

This drawing section covers different techniques. Activities 3.5a to 3.5c are suitable for resistant materials and 3.6a to 3.6c are for non-resistant materials.

3.5 DRAWING TECHNIQUES FOR RESISTANT MATERIALS

When drawing objects made from resistant materials – wood, metal and plastics – designers use a range of different techniques. Each is used for different purposes. Visualisations are quick drawings without detail (see pages 29–30 and 59–61 of your Nelson textbook) and don't need to be accurate. They can loosely follow the form of technical drawings. Technical drawings are used to convey a more accurate depiction of size, shape and placement. They have complex rules. If you are familiar with these rules – use them! In the following pages they have been simplified.

DESIGN FUNDAMENTALS 37

Suitable 3D drawings to use as visualisations (quick and not necessarily accurate) or design options (more refined) include:
- two-point perspective drawing
- oblique – cabinet and cavalier
- isometric drawings.

The 2D technical drawing technique most suitable for working drawings is:
- orthogonal drawing.

(Refer to the table on page 60 of your Nelson textbook to see when each drawing technique is appropriately used.)

> **TIP**
>
> **Technical drawing tips**
> You need a range of pencils when you are drawing in different styles:
> - 2B–6B pencils are softer and give you darker, thicker lines that are usually less accurate and tend to smudge – great for quick sketches.
> - HB pencils are harder and give a grey line that is thin and accurate – good for technical drawing.
> - Fineliners can be used to define your lines cleanly (once you have 'roughed out' the idea in pencil).
>
> Always keep your pencils sharp, particularly when drawing accurate technical drawings. Keep your guidelines light, so that they can be erased easily if they need to be changed or removed.

Activity 3.5a: Two-point perspective drawing

This style gives a realistic impression of the product. It shows the object on an angle so you can see two sides. Use the step-by-step instructions that follow to draw two-point perspective drawings of the product you are redeveloping in Unit 1, your design ideas or a design option (as a presentation drawing).

You can do these drawings freehand (for visualisations) or use a ruler (or drawing board and equipment) to achieve a more refined, presentation-style drawing.

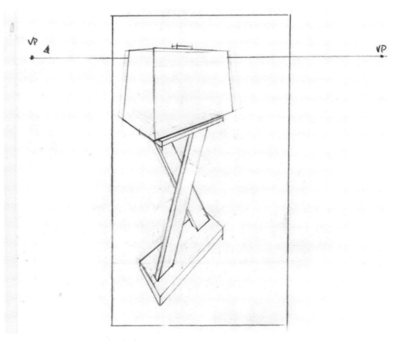

A lamp (left) and a two-point perspective drawing of the lamp (right)

9780170400404

DRAWING A SHELVING UNIT

1. To set up your page, draw a horizontal line (HL) across the top one-third. This is your horizon line (see box at right).
2. Draw two dots, one at each end of the horizon line. These are your vanishing points (VP). Label your vanishing points and your horizon line (HL), as shown in the diagram.

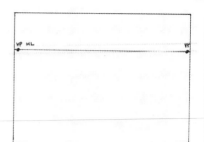

Vanishing points (VP)
You decide how far apart the vanishing points are: the closer together they are, the more exaggerated the sense of perspective is. If you want to make your drawing less exaggerated, add pieces of paper to the sides of your drawing page to increase the distance between your vanishing points. VPs are marked with dots.

Horizon line (HL)
The horizon line is usually placed somewhere around the top third of the page. It indicates your eye level. When the horizon line is high on the page, it is as though you are looking down on the object – like a bird's-eye view. Conversely, if it is low on the page, it seems as if you are on the ground, looking up at the object – like a worm's-eye view. HL is marked by a symbol resembling an eyeball (at right).

3. To draw the shelving unit (or your object), as in Step 6 you will begin with a box. The starting point for the box is a vertical line in the bottom two-thirds of the page. This is the front corner of the box. (Draw the vertical line to the left or right of centre to show more of the front.) Draw faint lines from the top of the vertical line to each of the vanishing points. Repeat from the bottom of the vertical line.

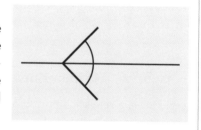

4. Decide where the back corners of the (shelving unit's) box are. Draw vertical lines between the angled lines to show this.

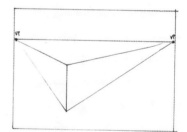

5. Draw two angled lines, one from the top of the left back corner of the box (left vertical line) to the right vanishing point, and the other from the right back corner of the box (right vertical line) to the left vanishing point. They should cross. Erase the lines that extend from the outside edges of the box to the vanishing points.

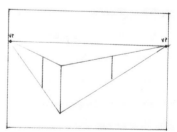

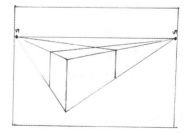

6. Now, draw in the details of the shelving unit (or your object), making sure that any horizontal lines are drawn towards a vanishing point and all vertical lines are drawn straight up-and-down. Use a ruler and square if necessary. To increase the sense of perspective, darken the lines around the front corner and leave the outside edges and back corner fainter. When you have finished your drawing, erase any guidelines. Outline the drawing in black pen for strong impact.

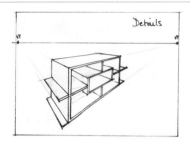

TIP

Points to note
- The sides of an object get smaller and any parts get closer together as they go towards the vanishing point. The halfway point of the side is found at the intersection of diagonal lines drawn (faintly) from each corner.
- Curves are distorted – some sections appear flatter and others appear tighter. You will need to research how to draw circles, curves and ellipses in technical drawings if they are part of your design.

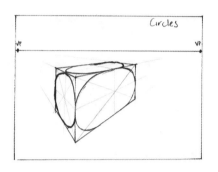

Activity 3.5b: Isometric and oblique drawing

Both of these forms of drawing give a sense of an object being three-dimensional, but are also drawn without the object getting smaller as it recedes into the background. As a result, the objects look strangely proportioned. However, they can be drawn to scale and accurate measurements can be taken from them. They can be a useful addition to a working drawing.

Both of these techniques are drawn over a grid (usually placed under your drawing page) or using a drawing board with T squares and 30° or 45° squares, so that all lines that recede in the drawing go back at a consistent angle from the base line.

Isometric drawing

In an isometric drawing, the horizontal lines on front and side faces recede at a 30° angle from the base of the page. All lines are drawn using scaled measurements. An isometric drawing is useful if you are drawing an object that doesn't have an obvious front, or if there are important details you want to show in both the front and side faces of the object. You can also take measurements directly from the drawing.

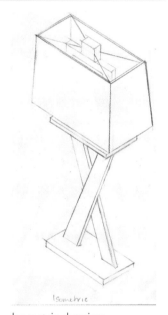

Isometric drawing

Oblique drawing

In an oblique drawing, the front face of the object is seen flat and face on, and the horizontal lines on the sides recede on a 45° angle. An oblique drawing technique is an appropriate style to use if your object has an obvious front and fewer details on the sides. There are two types of oblique drawing that are commonly used:
- **cavalier oblique** – where the receding lines are drawn to scale, i.e. the same scale as lines on the front face (so, if you were drawing a cube, all the lines drawn would have exactly the same length). As with isometric drawings, measurements can be taken directly from the drawing
- **cabinet oblique** – where the receding horizontal lines are drawn at half the scale of the front face of the object (this looks more realistic).

Cavalier oblique drawing

Activity 3.5c: Design options in isometric or oblique

To create true scale drawings, you will need to include measurements. Use a drawing board with instruments, appropriate grid paper to assist, or **computer-aided design (CAD)**.

Use a simple scale – 1:2, 1:5, 1:10, or if the object is small, 2:1. See 'Drawing to scale' on page 44.

Indicate the major dimensions of the object on your drawing. If there are hidden or complex details, show these as a larger detail drawn to the side of the main drawing.

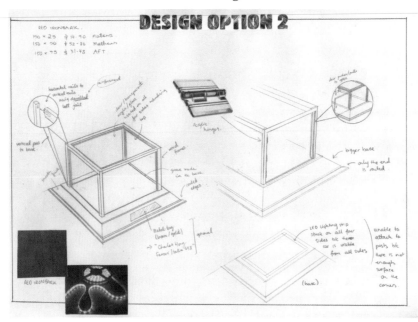

Design option by student Mei Hong with details in circles linked to the drawing

3.6 DRAWING TECHNIQUES FOR NON-RESISTANT MATERIALS

There is a range of styles and techniques for drawing non-resistant materials – fabric products and clothes.

Suitable 3D drawings to use as visualisations or design options include:
- fashion illustrations
- descriptive drawings.

Suitable 2D drawings for working drawings (see page 46) are:
- flats or trade sketches
- construction diagrams and pattern modifications.

Refer to the table on page 24 to see when each drawing technique is appropriately used.

Activity 3.6a: Proportions and using a croquis

First, you need to think about fashion drawing proportions. The real-life proportion of an adult body is that the body is about 7.5 times the height of the head. For fashion drawing, the figure that forms the basis of the drawing is called a **croquis** and is drawn at least 8.5–9 times the height of the head. For some more exaggerated stylised drawings, the distance below the knee is even further elongated.

1. Take an A4 sheet of paper and measure 2 centimetres up from the bottom of the page. Divide the rest of the page into eight even parts (easily done by folding).
2. In the top section, draw an oval for the head. The base of the neck is positioned halfway down to the next fold-line. Draw a strong horizontal line to indicate the shoulder line. The next fold-line positions the bust or centre chest line, and the following one, the navel. Draw a narrowing shape from the shoulder to the navel to give definition to the torso.

3   The middle fold is the position of the crotch or the widest part of the hips. Draw a wedge between the navel and the hip line. Two folds down is the position of the knees and a further two folds down, the last fold, shows the top of the foot, the ankle. Draw simple wedges below the last line for the feet, in the 2 centimetres allowed at the base of the page.
4   Play around with changing the angles of the strong horizontal lines at the shoulder, waist and hips to place the body in different positions, to twist and bend the form and to provide movement in your drawings.

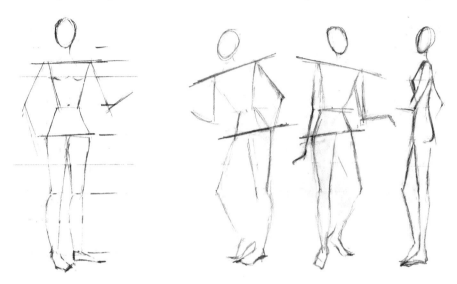

Folded drawings using the 8.5 fashion proportion (body = 8.5 lengths)

Drawing children

The proportions of children are very different to those of adults – the heads are a much bigger proportion of the body height. For example, a toddler's figure might be 4 or 4.5 times the size of the head, and a young teenager might have a ratio of 6:1 or 7:1.

Activity 3.6b: Fashion illustrations

In this activity, you will explore the use of line, shape, colour, texture and proportion, and experiment with different drawing media.
1   Choose 2–4 photographs of very different types of garments from fashion magazines. (If your focus is on other products, you can still follow the same process.) Try to find examples that show the figures in strong positions. Identify in each the most important or striking aspect of the garment design – line, shape, colour, texture, decorative detail, etc.
2   Create your own 'drawings' from the photographs. Explore different ways of highlighting the design feature(s) of each garment using different drawing or painting media. You could use: watercolours or watercolour pencils, oils, pastels, conté, coloured and metallic pencils, fineliners, broad- and fine-tipped textas, charcoal, cut or torn paper (in different colours, textures and patterns). Use a combination of two or three different materials, particularly ones that provide contrast (e.g. watercolour wash over pastel to give a sense of floaty, gauze-like fabric over something more solid).
3   When exploring different ways of portraying these designs, don't worry about details – simplify and stylise the face, hands, feet, etc. to one or two lines. Exaggerate parts of the body and the movement of the figure – particularly if they highlight the design features more – but still keep a rough sense of the body proportions explored in the previous activity.

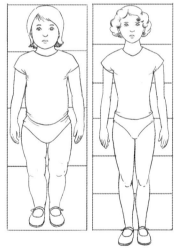

Proportions of children (a toddler and a teenager)

4 Find books that show a range of fashion illustrations. Look at the ways in which designers use drawing media, the design elements and principles (line, shape, colour, texture, balance, symmetry, proportion, positive/negative space, contrast, etc.), and how they portray the body in exaggerated and/or stylised forms. Try to apply the techniques you see to your drawings.

Illustration, by student Leya Mackus, showing that fabrics and paper can be used to demonstrate texture

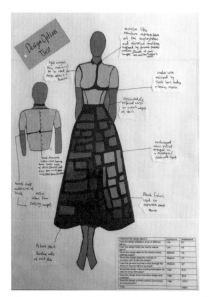

Design option drawings by students Chelsea Jones (left) and Madelaine Porrit (right)

Activity 3.6c: Descriptive drawings

Descriptive drawings are suitable for design option drawings (they are considered presentation drawings). They contain lots of detail – of the structure of the garments, indications of stitching, shape of collars, cuffs, bands, gathering, indications of the fall and weight of fabric, patterning, etc. (though not all details have to be shown in full). The outfits are usually shown in both front and back view.

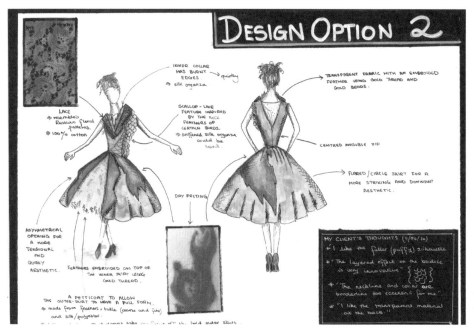

Annotated descriptive drawing, as a design option by student Lian Wilson

1. As a practice run, choose an outfit in your wardrobe – think about the whole 'look', including shoes, tights/leggings, hat, etc. Then follow the same steps to create a design option, using the best idea from your visualisations.
2. Using a figure croquis (or template shapes) choose and trace over an appropriate front and back view.
3. Draw your clothing onto those figures. Do this first with a greylead pencil, then go over the main lines and features with a fineliner. Add colour and texture with coloured pencils, pastel, and/or watercolour paint washes. Make sure you indicate the colour and patterning of the fabric, how it folds and drapes over the body, any obvious stitching, design features such as pockets, collars, bands, and show some indication of fastenings (buttons, zips, etc.).
4. Think about the presentation of your drawing. Experiment with different backgrounds. Cut out the drawings you have done (perhaps with a little white around the edges so that they stand out), then lay the drawings on different backgrounds – papers of different colours, textures and patterns, photos – and notice which highlights your drawing best. Be careful not to overdo this so that your drawing 'disappears'.

3.7 WORKING DRAWINGS

Working drawings for resistant materials

An orthogonal is the most suitable working drawing for resistant materials. It will give you the exact dimensions and placements of all parts. It can be done as a CAD drawing, or drawn up by hand with instruments, as follows.

Activity 3.7a: Orthogonal drawing

An orthogonal drawing is a two-dimensional drawing of several views of an object (usually front, top and side views) that are accurately aligned. It gives the details for construction. In Australia, we use the third angle projection method for orthogonal drawings.

Use the instructions that follow to complete an orthogonal drawing of your preferred option – you will need to calculate all the measurements to do this accurately.

The process outlined below is a simplified version of Australian Standards for orthogonal drawing. For more detailed instructions, download the document 'Technical Drawing specifications' from VCE Visual Communications Design, go to: VCAA VCE Study Design index, Visual Communication Design, Support Material section.

Setting up your page

A technical drawing has conventions of layout and presentation as follows:
1. Draw a border around your page – a single line 10 mm from the edge. If your object is tall and narrow, use a portrait layout, if your object is wide, turn your page to a landscape orientation.
2. Add a title block. The title block is drawn all or part of the way across the bottom of your page and divided into sections. It includes these details:
   - the title of your work (name of object)
   - the scale used
   - the type of drawing
   - your name and the date.
3. Draw a quick measured sketch of your object and note the major dimensions (length, width, depth) to help determine the best scale. Choose your scale to allow (at least) three views on an A3 page (approximately a quarter-page for each view). See the box 'Drawing to scale'.

> **TIP**
>
> ### Drawing to scale
>
> When indicating scale, the unit of the line used on your page is always written first, followed by the unit of measurement for your object (its real-life size). For example, in a scale of 1:5 (we say 'one to five'), every millimetre that you draw on your page is equal to 5 millimetres on the real object. So, if you draw a line that is 120 mm long, it will mean that the related section of the object is 600 millimetres long. The recommended scales of 1:2, 1:5 or 1:10 are suitable for most furniture or large objects. For smaller objects, such as a decorative box, use a scale of 1:2. For a very small object (e.g. a piece of jewellery), make your drawing larger than the object and use a scale of 2:1.
>
> Draw up a table with the actual measurements and their scaled equivalent for reference before starting, as shown in the table below for three scales. Add all the actual measurements of the object (in the first column) and calculate the drawing measurements at the appropriate scale.
>
> | Measurements of the object | Scale 1:2 | Scale 1:5 | Scale 1:10 |
> | --- | --- | --- | --- |
> | Length: 600 mm | 300 mm | 120 mm | 60 mm |
> | Height: 300 mm | 150 mm | 60 mm | 30 mm |
> | Width: 250 mm | 125 mm | 50 mm | 25 mm |
> | Thickness: 20 mm | 10 mm | 4 mm | 2 mm |

Drawing your object

4   Roughly plan where your three views are going to be placed on the page – think of your page divided into quarters and place a view in three sections. If the two side views are different, you need to show both (and your front view should be in the centre, between them).

5   Start with your front view (called 'elevation' in a building or large structure). Lightly draw a guide box in the scaled height and length dimensions of the object. Allow some space around the object for the view name and for dimension indications. Draw the outline shape of the front view and then add details, using accurate measurements. Make sure you draw in the thickness of the materials. Label the view and add the required dimension indications. Wherever possible, dimensions are placed outside the outline and always written in the same direction. Unlike in the drawing below, the longest dimensions are usually placed furthest away from the object.

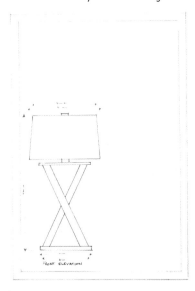

Front view

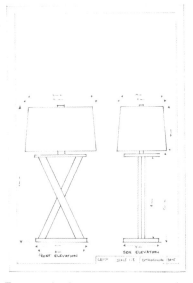

Front and side view

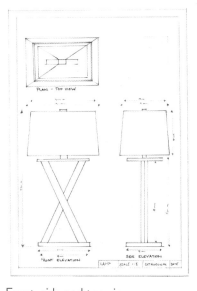

Front, side and top view

6. Next, tackle the side view or elevation (either left or right). To line up your side and front views, draw light horizontal guidelines that extend from the corners of your front view across to where your side views will be placed. Again, draw in a faint guide box using your scaled dimensions. Draw the outline of the side view and fill in any details. Make sure you include indications of joins, screws, nuts/bolts, etc.
7. Your top view (plan) needs to be placed directly above the front view. Use faint guidelines extended up from the outer edges of your front view. Place the top view the same distance away from the front view as the side view. Accurately draw in your outline and details. Label and add significant dimensions. To make your drawing clear, go over all outlines of the object with a black pen.
8. To add hidden details, see below.

> **TIP**
>
> **Dimensions**
> An orthogonal drawing has dimensions, which are used for construction. A few significant dimensions are important when sketching ideas, but accurate information is required on your final working drawing for a materials cutting list. It's important to read more detailed instructions on how to dimension in a technical drawing book or the VCE Visual Communication Design 'Technical Drawing Specifications' sheet.

Details

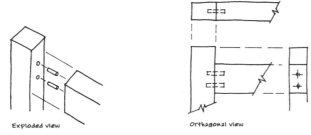

Exploded and orthogonal drawing of a dowel joint

Hidden details

Details that cannot be seen from the outside of the object – e.g. joins, drawers, internal shaping or components that may be inside or underneath an object – need to be included on your drawings. Showing hidden details helps with production planning. These details are drawn using a thin, dashed line.

Detail drawing

If you have details that are too complex to draw on a small or scaled drawing, draw these details to a larger scale in a circle (or box) on the side. Circle the area you are enlarging and connect the two circles with a line for clear communication.

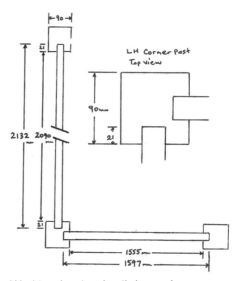

Working drawing details by student Jonathan Scampton

Working drawings for non-resistant materials

A 'flat' or 'trade sketch' is the most suitable working drawing for non-resistant materials. This type of drawing can be computer-aided; many designers use Adobe Illustrator. You can find many instructional videos on YouTube to help you. To draw by hand, use the instructions in Activity 3.7b.

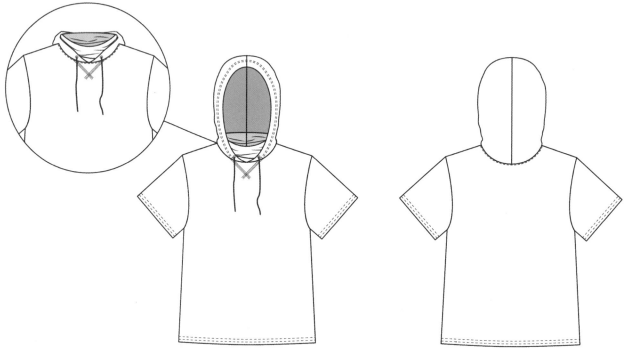

A flat drawing with detail, created using Illustrator by Nicole Crozier.

Activity 3.7b: Drawing flats

Flats are drawings used in the fashion/textiles industry to clearly explain detailed information to the manufacturer. Flats form a crucial part of a spec (specification) sheet, which also includes details of construction, sizing, fabric type and notions (e.g. buttons, zips, trimmings, ribbing). In flats, clothing is drawn off the body (laid down 'flat') and shows all details. They include expanded drawings of details. Flats normally show a front and back view of each garment (with one arm out and one folded down as shown in the drawing below), and aren't usually coloured.

1. To practise, choose one article of clothing from your wardrobe. Lay it flat and, using pencil, draw it exactly as you see it. Otherwise, imagine your preferred option design is lying flat on a table so that you are viewing all the details from either the front or the back.
2. Once you have correctly captured the shape and the main design features of the article of clothing, go over your drawing with fineliner (removing your pencil lines). Then add details of fold-lines, fabric texture, top-stitching, buttons, zips, pockets, bands/ribbing, collars and decorative details.
3. Next, draw expanded or 'exploded' drawings of any hidden details or aspects. Include annotation if needed to make the communication clear.

Making a model

It may be useful for you to make a scale model from paper, fabric, plastic, cardboard or plywood and glue to check the proportions of your design. You can make a model in the same scale as your working drawing.

Student Justin Cauchi building a model on a scaled working drawing

Design thinking strategies in your folio

Throughout your folio, you need to use a range of critical and creative thinking techniques. Here is a loose listing of the tasks that match each.

Creative thinking	Critical thinking
• Brainstorming • Use of graphic organisers • Exploration of design elements and principles • Developing design ideas (visualisations and design options) with SCAMPER or trigger words. • Incorporating end-user feedback and input • Combining ideas in unexpected ways • Suggesting new ways to use materials and processes • Coming up with something completely new and different, unlike the 'normal'	• Use of graphic organisers to make decisions • Defining the design situation through constraints and considerations • Seeking, interpreting and implementing end-user feedback and input • Developing evaluation criteria • Researching to find and check accuracy of information • Annotating design ideas to reflect your judgement on their feasibility and/or usefulness • Selecting the preferred option • Material testing and process trialing • Production plan decisions • Analysing the finished product using evaluation criteria

UNIT 1

SUSTAINABLE PRODUCT REDEVELOPMENT

This unit will cover the following areas in detail:

Area of Study 1: Sustainable redevelopment of a product

4	Starting Unit 1	p. 50
5	Product analysis: Investigating and defining	p. 54
6	Design brief and evaluation criteria	p. 62
7	Research, design and development	p. 64
8	Planning for production	p. 66

Area of Study 2: Producing and evaluating a redeveloped product

9	Production and recording progress	p. 68
10	Evaluation of the product	p. 69
	Assessment for Unit 1	p. 72

UNIT 1 CALENDAR

Week	Topic	Activities	Due date
1	Unit intro and overview **Outcome 1** Product design process Product design factors (sustainability and legal responsibilities)	1.1 The product design process 1.2 Product design factors and sustainability practices 4.1 Mini case study 4.2 IP or intellectual property	
2	Product selection and analysis	5.1 PMI of original product 5.2a Analysis with product design factors 5.2b Annotated analysis 5.3 Creative and critical thinking	
3	Design brief – outline of context and constraints and considerations Evaluation criteria	6.1 Design brief 6.2 Evaluation criteria	
4	Research planning Materials – categories, subcategories, characteristics and properties, appropriate uses; Materials testing; Process trials	2.3 Research planning 2.4 Research for inspiration 2.5–2.6 Materials research, test, process trials 2.7 Materials and sustainability	
5 6 7	Design and development • Creative and critical thinking (design elements and principles, how to be inspired etc.) • Drawing techniques • Selecting the preferred option • Justification of preferred option • Working drawings	3.1–3.3 Types of drawings 7.1 Visualisations 7.2 Design options 3.4 Creative drawing techniques 3.5 Drawing for resistant materials 3.6 Drawing for non-resistant materials 7.3 Selecting and justifying preferred option 7.4 Working drawings	
8	Production planning components Safety – identifying hazards, assessing risk and risk control	8.1–8.3 Steps, timeline and materials for redeveloped product 8.4 Risk assessment for product and processes 8.5 Quality measures	
9–14	**Outcome 2** Production – making the product	9.2 Journal and record of changes to plan	
15	Product comparison and evaluation Testing of product and comparing with original design	10.1–10.2 Product evaluation report 10.3 (Optional) Evaluation of process	
16	Finishing off all loose ends		

CHAPTER 4 | Starting Unit 1

Using the product design process and product design factors
4.1 Mini case study..p. 50
4.2 IP or intellectual property...p. 52

To become familiar with the product design process and the product design factors, read and complete the activities in 1.1 and 1.2.

4.1 MINI CASE STUDY

Activity 4.1: Designers and sustainability practices

In Unit 1 you are expected to redevelop a product, change it significantly and improve its level of sustainability.

Looking at the way professional designers incorporate sustainability strategies can give you ideas for this.

Select one or two designers who work in a similar field or with similar material to you and prepare a brief case study for each designer. Choose someone you can find lots of information about, particularly regarding the materials they use or their choices during production and throughout the supply chain. Possible sources of information include magazine articles, websites, exhibition catalogues and design books.

There are many suitable case studies of designers in your Nelson textbook – look in the table of contents. When analysing the sustainability practices of your selected designers, refer to the general environmental design strategies (page 124) as well as strategies for specific materials. Examples: Bodypeace, Schamburg + Alvisse, Cantilever, Etiko and Büro North.

You could present your case study as a class talk, poster or multimedia presentation (e.g. Word document with text and images, PowerPoint, Prezi, web page).

If you are looking at two designers, then repeat this activity using the same questions below.

1 Name of the designer: _____

2 What do they design and make?
   _____

3 How does the designer approach the following areas to incorporate sustainability strategies in their product?
   - **Design**
     How does the design of the product impact on its level of sustainability? How does the designer try to minimise the impact of the product through its design?

     _____
     _____
     _____
     _____

- **Materials**

  What are they? What are the raw forms of these materials? Where (location) do they come from? How are they converted into useable materials? What happens to these materials when the product is discarded? Why are they considered sustainable? Do any of the materials have sustainable accreditation?

  _____
  _____
  _____
  _____

- **Production, distribution, use and disposal**

  How does the designer claim to be sustainable in these areas?

  How and where is the product made? What information is available on the treatment of workers, negative health impacts, toxic by-products, etc. during production? How far does the product travel from factory to retail and how is it transported?

  How durable is the product? Does it last for a longer or shorter time than similar products? Can it be repaired? Does its maintenance use up lots of resources (e.g. when cleaning) or create waste (e.g. microplastics)? Does it have clear instructions on how to use and care for it? What can be done with the product when it is discarded?

  _____
  _____
  _____
  _____
  _____

Social responsibility

4 How does the designer's product benefit the people who buy it (consumers) and society in general?

_____
_____
_____
_____

5 If you looked at two designers, make a statement that sums up which designer's practices you think are more sustainable and why.

_____
_____
_____
_____

6 Discuss ways that you think this designer/s can prove their claims of having sustainable practices.

_____
_____

4.2 IP OR INTELLECTUAL PROPERTY (FOR ALL UNITS)

Intellectual property (IP) is the property of the mind. There are different protections that designers can rely on to make sure that other people do not benefit commercially from their ideas. Each form of protection is appropriate for different circumstances.

Find out about:
- copyright
- patents
- design registration.

(For information on intellectual property, see pages 108–112 of your Nelson textbook.)

Activity 4.2a: Forms of intellectual property protection

The following table is muddled. Fill in the blank table on the next page with the information in the correct places.

Type of protection	What it applies to	Length of protection	Paperwork needed	What it protects	Cost
Patent	A design intended or used for mass production that has not been shown publicly	Over 50 years	None, it applies automatically	Technical details	$250–$700
Copyright	Technical or mechanical information relating to a new invention or new way of making a product work	20 years	A design must be registered	The visual appearance of a product	$5000 to $10 000
Design registration	Writing or two-dimensional drawing/design (or a one-off object based from a design)	10 years	An international search is done to see if the idea is new/unique, then the patent is applied for	Work that expresses an original idea	Free

Type of protection	What it applies to	Length of protection	Paperwork needed	What it protects	Cost
Patent					
Copyright					
Design registration					

Activity 4.2b: Examples of intellectual property protection

Go to the IP Australia website and find an image of a product for each form of IP in the table. Insert the image into a document and annotate it to explain which aspect of the product is being protected.

Activity 4.2c: Attributing intellectual property

Unit 1 requires you to redevelop a product, i.e. use someone else's ideas as the basis for your designs, but not to copy. You need to honour the moral rights of the designer by attributing their work in a clear and obvious way to the viewer.

An easy way to be sure you know the name of the creator is to start by looking at the work of a particular designer or company and choose one of their products, rather than start with an image and try to find the creator's name. In some cases, you may only be able to find out the name of a manufacturer or retailer. Keep in mind that Google Images and Pinterest are not the creators of works; they are only equivalent to online libraries or galleries.

After reading the above, explain, in point form, what you can do to:
- be sure of the name of a product's creator
- attribute their work clearly
- ensure you are not copying the idea too closely.

CHAPTER 5
Product analysis: Investigating and defining

Unit 1, Outcome 1
5.1 Selecting your product .... p. 54
5.2 Analysing the original product .... p. 56
5.3 Creative and critical thinking .... p. 59

In Unit 1 you are required to redevelop an existing product. The first steps in the product design process will be:
1. selecting and analysing a product and defining its redevelopment
2. writing the design brief
3. creating **evaluation criteria**
4. carrying out research.

5.1 SELECTING YOUR PRODUCT

Select an existing product to redevelop (your teacher may give you guidelines for this). You can maintain the primary function of the original product but must improve the product in several ways and in its sustainability. When looking for a product, choose one that you know will have plenty of information available, such as:
- the name of the creator or the company
- materials used
- details of any components
- details of its appearance from several views, including any decorative details
- when it might be used and for what purpose
- construction details.

If you don't have the actual product, find images of the product that provide you with lots of detail. A low definition image from the Internet without supporting information might not be suitable. If you can't get a photo, draw the product in as much detail as possible, with notes explaining the product's features.

For the purposes of clarity, the product to be redeveloped will be called the original product and the new product you design will be called the redeveloped product. The following activities will help you analyse the original product. Check with your teacher which activities you need to complete.

Activity 5.1: PMI analysis of the original product

Use this page to note down information for the analysis of the original product. Use the questions from Activity 1.3 to assist you.

A3

Image of product
Photograph or detailed drawing

Description:
- Size _____
- Colour(s) _____
- Materials _____
- Fittings _____
- Quality _____
- Joins _____

Plus
Note the good things about the product.

Minus
Write down the things about the product that are not good, or don't work well, such as the materials, the components, the way it looks, its size, etc.

Interesting
What is interesting, unusual, or gives you ideas for changing this product?

Sustainability
Comment on the sustainability of this product and/or suggest improvements.

5.2 ANALYSING THE ORIGINAL PRODUCT

While you will not need to 'redevelop' every product design factor in your original product, a thorough analysis will help you select the most appropriate aspects to change. In the following activity you are provided with a question for five factors for both the original and the redeveloped product.

Activity 5.2a: Analysis using the product design factors

Purpose, function and context

The purpose is the reason the product exists. The function is what it does and how it works. Products have a primary function and secondary functions (aspects of the product that work together). The context includes how the product will be used, where, when and who will use it.

(For information on primary and secondary functions, see pages 92–3 of your Nelson textbook.)

Some questions to ask of the original product

What is the original product's primary function? What secondary functions support the primary function? In what context is the product used?

_____

_____

For your redevelopment

What will be the context for your redeveloped product? (e.g. a different location or event)
What functional aspects could be improved for this redevelopment?

_____

_____

User-centred design

This relates to how well the product is suited to the user; the human body and how the body works (**ergonomics**), their age, mobility or culture. Designers often use **anthropometric** data (average measurements of the body) to help them design products to suit most people or for specific sizings. Safety is closely related to ergonomics – if a product is not properly suited to the body and does not allow for how the body moves, the user's body could be strained or injured.

(For information on ergonomics and specific anthropometric data for adults and children, see pages 101–2 of your Nelson textbook or refer to the textbook's anthropometric data charts, which are available on the student book website.)

Some questions to ask of the original product

Who is the typical end-user of the original product? What are the ergonomic and/or safety aspects of this product? What anthropometric data or particular measurements could have been useful to know for this (original) product?

_____

_____

For your redevelopment

Who will be the typical end-user of your redeveloped product? What safety, ergonomic or comfort issues do you need to think about or could improve when redeveloping this product? Do they have any special needs or requirements?

_____

_____

Visual, tactile and aesthetic factors

Designers use design elements and principles thoughtfully to create **aesthetic** appeal in their designs and to analyse existing products. The design elements are the basic building blocks of design. The design principles refer to how these elements are put together, organised and manipulated.

Design elements		Design principles	
Line Shape Form Point Colour	Tone Texture Transparency, translucency and opacity	Proportion Pattern Movement/ rhythm Repetition and pattern Balance	Positive/ negative space Symmetry/ asymmetry Contrast Surface qualities Emphasis

(For information on design elements and principles, and specific information about how they are used in products, see pages 74–89 of your Nelson textbook.) Complete some of the design activities on pages 30–6 of this book to become familiar with this language.

Some questions to ask of the original product

How are the design elements and principles used in the original product (e.g. thick red straight lines)? What are the tactile characteristics, i.e. how does it feel to touch? Do you find it visually or aesthetically pleasing?

_____

_____

For your redevelopment

How could the visual, tactile and aesthetic characteristics of the product be improved in the redevelopment?

_____

_____

Materials used and sustainability

It is important to know the materials the original product is made from, particularly as you are required to improve its sustainability. Material choice contributes greatly to the environmental sustainability of a product; this also applies to the material used in the parts and components.

(See Chapter 5 and pages 163, 182–3, 197 and 216 of your Nelson textbook for information on identifying materials.)

Some questions to ask of the original product

What are the main materials the original product is made from? What characteristic/s or property/ies of the material/s are important for this product? Research the sustainability issues related to these materials, e.g. the source, the raw form, details of processing, life (or longevity) and how to dispose of the material/s and the components. How are the components attached (permanent or removable), and how do they impact on the recycling or disposal of the product?

_____

_____

For your redevelopment

How could the choice of materials in the redeveloped product improve its level of sustainability?

_____

_____

Analysis of quality

Aspects of quality in products relate to the choice of material used; construction methods; quality and placement of components; how it performs; and its visual design, e.g. the way it is proportioned; effective use of contrasting materials; decorative details and whether it exhibits classic or cutting-edge style.

A well-constructed, beautiful-looking product of suitable material with well-chosen components can not only be durable but will be cherished by its owners – and will therefore be long lasting and sustainable.

Some questions to ask of the original product

What is the quality of the original product in terms of materials, construction, components, etc.? Do you consider this product to be of high quality? Explain.

_____

_____

For your redevelopment

How can you improve the quality of the product through your redevelopment?

_____

Activity 5.2b: Annotated analysis of the original product

Take an image of the original design and attach it to the centre of a folio page. Take the information you gathered in Activity 5.2a and annotate the original product, covering all those factors and any others that are relevant. Use arrows as in the following examples. Use several pages, each with the image or parts of it, if you need more room. Alternatively, attach a see-through sheet over the image (either tracing paper, baking paper or use the plastic pocket) and write on that.

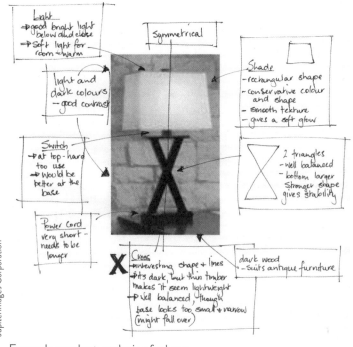

Example product analysis of a lamp

Analysis of an exercise bag

5.3 CREATIVE AND CRITICAL THINKING

Critical thinking means looking at something objectively and from an informed viewpoint in order to make decisions. It goes hand in hand with creativity and is important as it leads towards decisions that are realistic, practical and possible to achieve. It is not a negative way of looking at something, but a constructive method of analysing 'good' and 'bad' features and thinking creatively about ways to improve them. State clearly what you don't like about the original design and be creative by suggesting alternatives.

(For more detail on critical thinking, see pages 53–5 of your Nelson textbook and the associated activities in the 'Design fundamentals' section of this book.)

Your redeveloped product needs to be significantly different and you need to improve its sustainability in comparison to the original.

Think about changing:
- the design of secondary functions
- visual, tactile and aesthetic aspects (using the design elements and principles)
- sustainability and quality of materials, construction, components and finish.

Activity 5.3: Exploring ideas for redevelopment

Develop a mind map or graphic organiser that identifies all of the possible areas you could redevelop or change. (Use the dot points above as areas for change.) Refer to your responses to Activities 5.2a and 5.2b, particularly where you have identified areas for improvement. This exploration will assist you in developing your design brief and when designing. Use the layout on page 61 or create your own similar to the examples below.

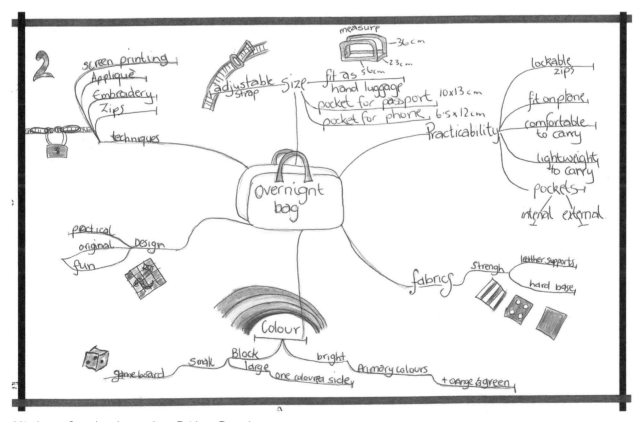

Mind map for a bag by student Bridget Bottcher

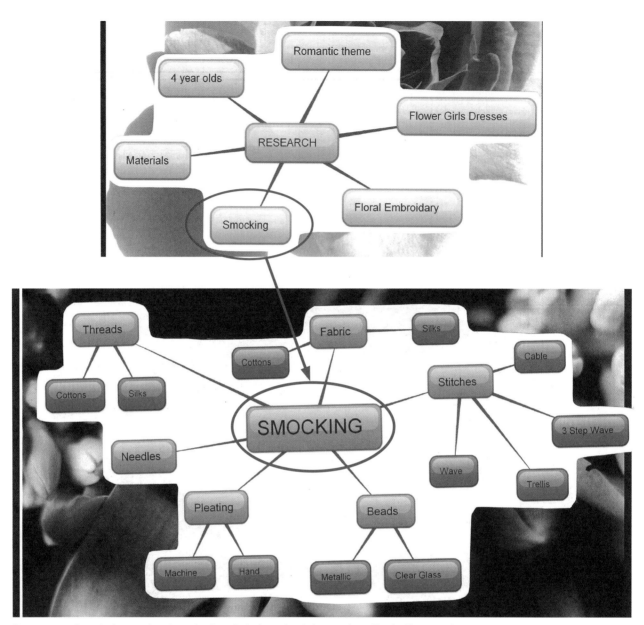

Segments of a mind map developed using digital methods by student Emily Fairweather

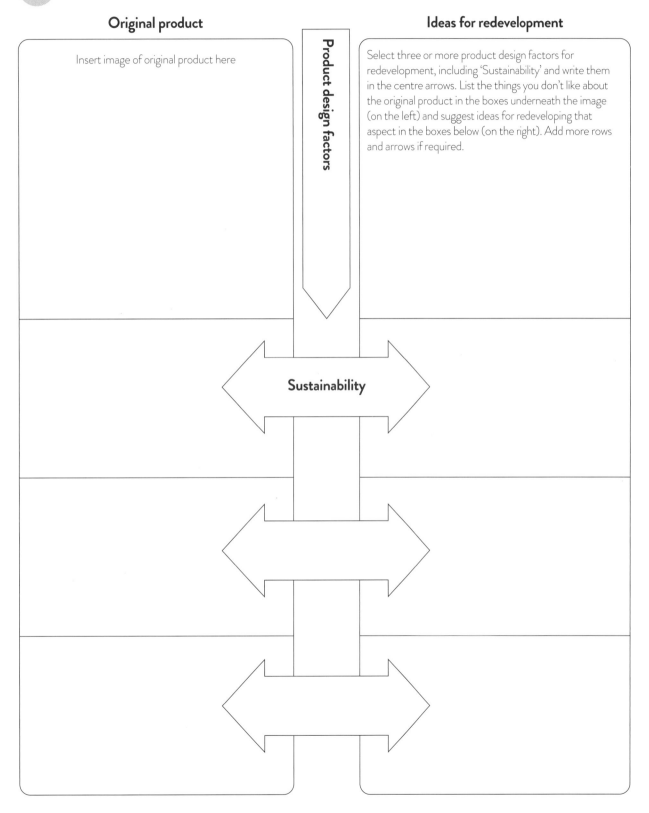

CHAPTER 6

Design brief and evaluation criteria

Unit 1, Outcome 1
6.1 Writing a design brief .... p. 62
6.2 Evaluation criteria .... p. 63

6.1 WRITING A DESIGN BRIEF

The design brief defines what is required, and acts as a point of reference during the design process. In redeveloping your product, you are required to solve a problem related to the original product, e.g. design a bag for a different purpose or a seat for a different user. This could also be described as maintaining the primary function, which is to carry things (for a bag) or to provide seating (for a seat).

Start by deciding on the specific purpose of your redeveloped product. Note that one of the constraints has already been given to you: improve the product's sustainability.

Use the design brief web activities from Activity 2.1 to plan the information required for your design brief.

Activity 6.1: Design brief

Use the following table for your design brief.

Outline of the context (from your design web) – include an explanation of the original product (i.e. its primary function, appearance and who it is designed for); the reason why it needs to be redeveloped and information about the user (relevant to your redevelopment).
The original product is a _____ designed by _____.
It is used for _____ and the typical end-user would be a person who: _____
I am redeveloping it so that _____ .

Constraints and considerations:
• Sustainability of the _____ (insert name of original product) must be improved.
• The redeveloped _____ must be finished by _____.
• The budget is $ _____ (or the material available is _____).
• _____
• _____

Changes to the original product will be needed in these areas:	These changes will improve the product because:
• _____	• _____
• _____	• _____

6.2 EVALUATION CRITERIA

A4

Read about evaluation criteria on pages 19–21 of your Nelson textbook and look at the sample questions in Chapter 2. Use them to help you to complete the table in Activity 6.2.

Activity 6.2: Evaluation criteria table

Develop at least five evaluation criteria from your design brief. Be sure to include one about the product's improved sustainability. In the second column, justify or explain why the criteria you have chosen are important. In the third column, suggest activities that could help you achieve the criterion, ranging from research to skill trials. In the fourth column, explain how each criterion can be checked or judged when the product is finished, i.e. a method of measuring or deciding on how the product has satisfied each criterion.

Evaluation criteria for redeveloped product (as questions)	Justification Explanation of importance to the design brief and user, reason for choice	How to achieve this (suggest research to help, ideas to explore, tests or trials to perform etc.)	How it might be checked or tested (or compared with the original)

Add more rows if necessary.

CHAPTER 7
Research, design and development

Unit 1, Outcome 2
7.1 Research and visualisations .... p. 64
7.2 Design options .... p. 64
7.3 Time to apply your critical thinking .... p. 65
7.4 Working drawings .... p. 65

7.1 RESEARCH AND VISUALISATIONS

Activity 7.1a Research

Before starting your Design and Development work, be sure to check with your teacher the research activities (Activities 2.3–2.7) you need to complete from the 'Design fundamentals' section.

You should start with research planning, and consider how to include the research you need for:
- inspiration
- your end-user
- learning about materials through research, tests and trials
- sustainability and ways to improve it in your redeveloped product.

Activity 7.1b Visualisations

Before you start, read the sections on visualisations in your Nelson textbook (on pages 29–30, 59–61 and 69) and in the 'Design fundamentals' section, Chapter 3. Use some of the ideas under 'Thinking and designing creatively' (3.4) to help you develop exciting and innovative design ideas.

Create your own visualisation pages (idea development) by referring to page 183 of the Design Folio Template. You can use 3D models or any of the drawing techniques (suitable to your material category) from Chapter 3. Although many of the drawings for resistant materials are technical drawings with rules, you can draw them quickly, roughly and using freehand in this step.

Draw some visualisations near images from your research to show how they influenced your ideas.

7.2 DESIGN OPTIONS

Read the sections on design options in your Nelson textbook (on pages 61–3 and 69–71) and in the 'Design fundamentals' section, Chapter 3. Use the detailed information about different drawing techniques to guide you in drawing your design options.

Activity 7.2: Design options

Use the layout for design options from the Design Folio Template (page 184) to develop two or three design options using the appropriate drawing techniques from Chapter 3.

7.3 TIME TO APPLY YOUR CRITICAL THINKING

Once you have several design options, you can apply your evaluation criteria, which is considered to be a 'critical thinking' technique. If you score each option against the criteria, you will have made some concrete decisions.

Selecting and justifying your preferred option

(For explanations on how to select the preferred option, see pages 32–3 of your Nelson textbook.)

You need to choose your preferred option from the design options developed. In Unit 1, remember to consider which option improves the original product's level of sustainability.

Activity 7.3a: Using a grid to select (optional)

Score each evaluation criterion out of five for each option. Total up the score for all the options as in the table on page 185 of the Design Folio Template. Extend or enlarge the table as necessary.

Show and discuss your designs with others (your teacher, class colleagues, friends and parents) for feedback to help you score them. This will boost your confidence in the decisions you have already made, and help you to clarify your decision and to write your justification.

Activity 7.3b: Justifying your preferred option

1 Which design option did you choose as your preferred option, i.e. which scored the highest?

_____

2 Explain why you have chosen this design. (Refer to your evaluation criteria. How does this design satisfy its purpose and improve the sustainability of the original product?)

_____
_____
_____
_____
_____

3 What feedback did you gather? Who did you talk to about your design options, and what did they think?

_____
_____

7.4 WORKING DRAWINGS

Read the Working drawings section of your Nelson textbook on pages 33–5, 63–7 and 71–3, and 3.1 in the 'Design fundamentals' section of this workbook. This will give you guidance to help you choose and apply the correct drawing techniques.

Activity 7.4: Developing working drawings

Use the working drawing page of the Design Folio Template (page 186). A working drawing should show at least two views of the preferred option, with all construction details clearly drawn, and with detailed measurements and information provided about the materials and components used.

CHAPTER 8 | Planning for production

Unit 1, Outcome 1
- 8.1 Materials list and costing .... p. 66
- 8.2 (Predicted) Production steps .... p. 66
- 8.3 Timeline .... p. 66
- 8.4 Risk assessment .... p. 67
- 8.5 Quality measures .... p. 67

The scheduled production plan

Your scheduled production plan requires five main components, which are listed above. If you choose to make a 'prototype' as your final product, you need to take into account the cutting, joining and finishing processes to be used and incorporate them into your planning. (Review the 'Planning' section of your Nelson textbook (pages 36–47) for details about what is required.)

Check with your teacher what is possible and acceptable as a prototype for assessment (this is also described on page 45 of your Nelson textbook).

8.1 MATERIALS LIST AND COSTING

Activity 8.1

Complete a materials list and costing (Design Folio Template, page 187) to calculate all the materials, components and other requirements you will need to make your product.

8.2 (PREDICTED) PRODUCTION STEPS

Activity 8.2

Complete a detailed step-by-step description of how you expect to make your product using the appropriate template in the Design Folio Template (page 188). Include safety and estimated time (for each step and a total). If using instructions from a commercial pattern, only include relevant ones. Think ahead about what will be involved in: measuring, marking out, cutting and/or shaping (or forming), the various joining methods and the steps involved, any embellishing or decorating, assembling and finishing.

8.3 TIMELINE

Activity 8.3

Using the Gantt chart format in the Design Folio Template (page 189), create a **timeline** that shows at what date (based on your workshop access) you expect to carry out each of the production steps.

Steps can be written very briefly in the timeline first, to work out the flow of production (the detailed description of steps (8.2) could be done later). It doesn't matter in what order you do 8.1 or 8.2 as long as the number of production steps is the same in each. Some steps may take several workshop sessions, while others can be combined in a single session.

8.4 RISK ASSESSMENT

Activity 8.4

Carefully read the section on risk assessment on pages 41–3 of your Nelson textbook. The areas you need to understand that relate to risk assessment include:
- **hazards** – the things, actions, behaviour or situations that are dangerous
- possible injuries
- calculating risk
- methods of control used to reduce incidents and injury.

Use the table provided in the Design Folio Template on pages 190–1 to complete your risk assessment and to assess the specific risks involved in:
- one aspect of pre-production
- at least three significant aspects of production (these could relate to machinery, materials, chemicals or specific processes)
- one aspect related to your work environment
- one aspect related to the use of your product.

Use the Safe Operating Procedures template on page 193 to document your training and competency on equipment identified in your Risk Assessment.

8.5 QUALITY MEASURES

Activity 8.5

Using the table in the Design Folio Template (page 192), create a list of at least three major processes in your production. List **quality measures** you could use for each process to help you achieve quality workmanship. You can add to this table as you learn new techniques.

CHAPTER 9

Production and recording progress

Unit 1, Outcome 2
9.1 Production .... p. 68
9.2 Documenting your progress .... p. 68

9.1 PRODUCTION

Apply risk management and quality measures to complete your production safely and accurately. Follow your plans and timeline to be as efficient as you can.

Refer to your quality measures for each step and the standard expected. During all these processes:
- practise using equipment
- trial processes to improve your skill level
- work accurately and with precision
- work with patience, particularly when finishing.

9.2 DOCUMENTING YOUR PROGRESS

Take photos at various stages of production and explain them. Your written log should outline:
- the date and time spent
- the processes you completed (as listed in your production steps)
- the materials, tools, equipment and machinery used
- what you have learnt, any difficulties you had, any construction problems you solved, any help you received and any changes to your designs and plans.

Looking back on the log will help you later, when writing your evaluation report. It will help you see where problems occurred and consider where improvements can be made. Some students also find it useful to write in their log what they need to do in the next session, particularly if it is a week away!

Activity 9.2: Your journal or log of work

Use the journal format in the Design Folio Template (page 195) as a basis for your journal or log of work. Create your own digitally to insert photos.

CHAPTER 10

Evaluation of the product

Unit 1, Outcome 2
10.1 Evaluating your finished product .... p. 69
10.2 Comparing your product to the original .... p. 69
10.3 Evaluating the process (optional extension activity) .... p. 71

10.1 EVALUATING YOUR FINISHED PRODUCT

Activity 10.1: Using your evaluation criteria

Use the table in the Design Folio Template on page 198 to evaluate your finished product. Make sure you:
- insert the evaluation criteria you wrote earlier
- explain how you tested or checked the finished product against each criterion using the method you suggested (in the four-part evaluation criteria)
- record your judgements and suggest improvements.
  Seek the opinions of others (e.g. teacher, friends, family) when needed.

DFT

10.2 COMPARING YOUR PRODUCT TO THE ORIGINAL

Activity 10.2a

How does your redeveloped product differ from the original design?
Detail the modifications you made, the materials used and how you have improved the original product's sustainability.

- _____
- _____
- _____

Compare the following relevant areas: (circle the relevant areas you redeveloped and respond below)

How well does your product function or work in comparison to the original?
How does the appearance and aesthetics of your product compare with the original?
How does the quality of your product compare with the original?

_____
_____
_____

Suggestions for further improvements that could be made to your product (include additional sustainability improvements).

_____
_____
_____
_____

Activity 10.2b: Visual comparison and evaluation

Complete an annotated visual comparison and evaluation of the original product and your redeveloped product; use arrows to indicate what your annotations refer to and how they can be seen in both. Be sure to point out the differences (numbering them clearly, for example, could help).

The following are suggestions for changes to be annotated.

Purpose, function and context Context and primary function; specific purpose of (a) original (b) redeveloped; secondary functions in each (both similar and different); size/operation/quality.	Visual, tactile and aesthetics Colours, patterns, lines, shapes, proportions, texture, contrast, etc.
Materials Name them, explain quality or more suited characteristics and properties for your redeveloped product.	Sustainability Comment on the sustainability of your redeveloped product compared with the original.

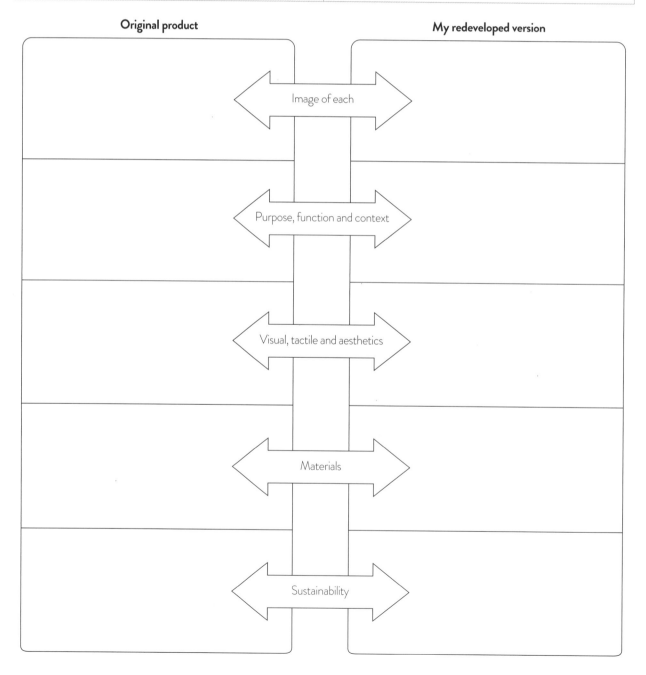

10.3 EVALUATING THE PROCESS (OPTIONAL EXTENSION ACTIVITY)

Activity 10.3: Evaluating your design, planning and production process

Use the following table as a guide for an in-depth evaluation of your design, planning and production process. This will help you to learn from your experience and be more effective and efficient next time. Ask your teacher which questions are to be answered. You can use this evaluation to help improve your process in Units 2, 3 and 4.

Design • Did you research widely and thoroughly explore the design and development stage to create a great solution? Explain. • Did you make changes to your design during production? Why?	
Working drawings • Were your working drawings accurate and clear, with enough detail for you to follow? Explain.	
Production plans • How useful was your scheduled production plan? Did you follow it? • Did you stick to the timeline? How did you use it? Did you have any problems with time? • Was your risk management sufficient to avoid any incidents?	
Production • Explain the positives and negatives of working with the materials you chose. • Explain the suitability and accuracy of the processes you completed to measure and mark out, cut, shape, join, assemble and/or embellish. • How could the processes be improved in quality, safety, accuracy and time taken?	

ASSESSMENT FOR UNIT 1

Outcome 1

Task	Expectations You need to:	Mark
Product design process, factors, sustainability and intellectual property	• identify the stages of the product design process • recognise the product design factors • understand intellectual property (IP) • describe the way a particular designer incorporates sustainability practices	/5
Original product analysis	• identify the designer/maker of the design • analyse and describe the original product (PMI, function, ergonomics, aesthetics, quality, materials etc.) • identify areas and ideas for redevelopment	/5
Design brief and evaluation criteria	• identify the design situation or problem clearly, with a range of suitable constraints and considerations • develop suitable evaluation criteria for the product and to select the preferred options	/10
Research	• identify areas of relevant research • identify material categories, classification, and properties and characteristics • annotate images of similar designs/products for inspiration • complete a materials test report • conduct trials of processes	/8
Design and development	• complete visualisations (including exercises with elements, principles, inspirational images and drawing styles) and models if appropriate • create three design options, apply the evaluation criteria to select the preferred option – and give justification • complete working and construction drawings	/12
Scheduled production plan	• complete a list of materials, components and fittings • write production steps for construction • create a timeline for these steps • complete a risk assessment • create measures to achieve quality workmanship	/10
	Total	/50

Outcome 2

Task	Expectations You need to:	Mark

Safe working during production	• work safely – apply risk management during production
	• complete the product within the time allowed and show:
	  – appropriate use of materials
	  – safe and competent use of equipment
	  – skill and accuracy in production processes
	  – high level of quality in all processes and finish stages
		/40

Evaluation	• evaluate finished redeveloped product using predetermined evaluation criteria • compare redeveloped product with original design	/10
	Total	/50

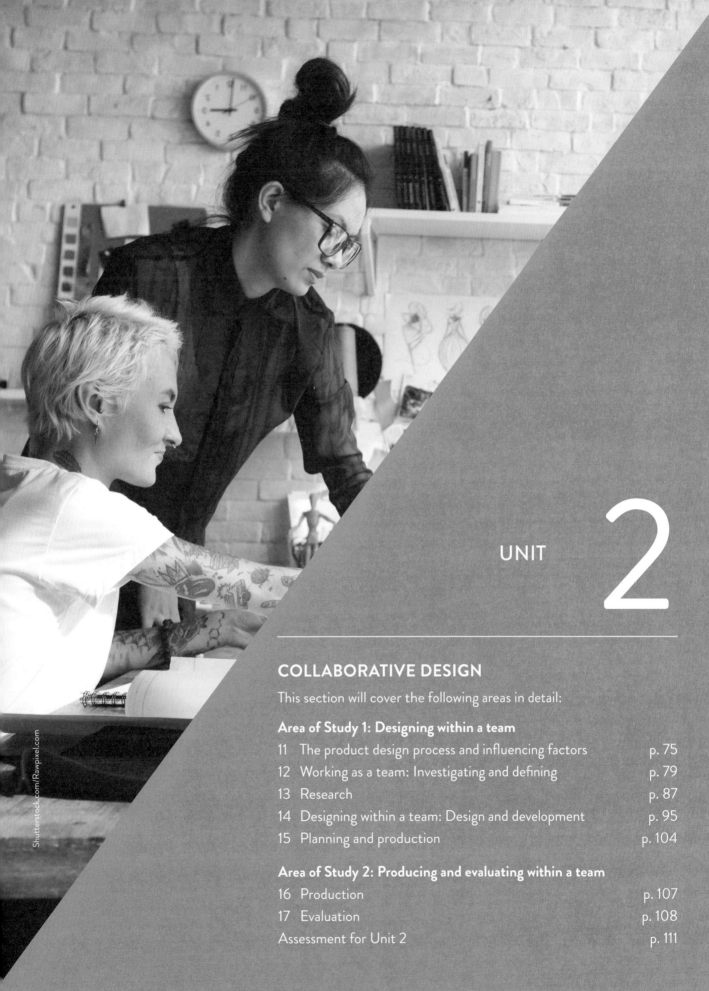

UNIT 2

COLLABORATIVE DESIGN

This section will cover the following areas in detail:

Area of Study 1: Designing within a team

11	The product design process and influencing factors	p. 75
12	Working as a team: Investigating and defining	p. 79
13	Research	p. 87
14	Designing within a team: Design and development	p. 95
15	Planning and production	p. 104

Area of Study 2: Producing and evaluating within a team

16	Production	p. 107
17	Evaluation	p. 108
Assessment for Unit 2		p. 111

UNIT 2 CALENDAR

Week	Topic	Activities	Due date
1	Intro and outlines **Outcome 1** Product design process, product design factors, user-centred design, sustainability	Review cross-study specifications in Chapter 1, complete activities as required 11.2, 11.3 and 12 Discussions, forming of teams	
2 3	Collaboration and ICT • Scenario, brainstorming • Design brief • Evaluation criteria Establishing the 'rules' and negotiating roles	Review Chapter 2, complete activities in 2.1 and 2.2 as required 12.1–12.2 Team roles 12.3–12.4 Design brief 12.5 Evaluation criteria	
4	Research from both primary and secondary sources • A design movement, culture or style for inspiration • Information, measurements, materials test, process trials	Complete activities in 2.3–2.7 as required 13.1–13.4 Research	
5 6 7 8	Design and development • Ideas, mood boards • Design options • Criteria grid, decision-making and justification • Working drawings	Review Chapter 3, and complete activities as required 14.1–14.3 Visualisations 14.4–14.5 Design options, feedback, selecting the preferred option and justification 14.6 Working and construction drawings	
9	Production planning • Steps for production (work plan) • Timeline • Materials list and costing • Risk assessment • Quality measures	15.1–15.2 Work plan 15.3 Timeline 15.4 Materials list and costing 15.5 Risk assessment 15.6 Quality measures	
10–14	**Outcome 2** Production	16.1–16.2 Journal and record of changes to plan Finished product	
15	Evaluation • Checking, feedback and discussion • Writing the report	17.1 Product, production and collaboration evaluation	
16	Presenting work to others	This could be done at any stage in the semester	

CHAPTER 11

The product design process and influencing factors

Unit 2, Outcome 1
11.1 Cross-study specifications .... p. 75
11.2 User-centred design .... p. 75
11.3 Presenting work to others .... p. 78

(For detailed and helpful information for this unit, refer to Chapter 10 of your Nelson textbook.)

11.1 CROSS-STUDY SPECIFICATIONS

There are a number of cross-study specifications already covered in the 'Design fundamentals' section, so if you are new to this subject or unfamiliar with these terms, familiarise yourself by completing the following activities:
- Activities 1.1a and 1.1b: The product design process, pages 2 and 3
- Activity 1.2a: The product design factors, page 3
- Activity 1.2b: Sustainability (one of the product design factors), page 5.

One of the product design factors that is emphasised in Unit 2 is the user-centred design approach. Some activities to help you understand this factor follow.

11.2 USER-CENTRED DESIGN

(For information on user-centred design, see pages 96–102 of your Nelson textbook.)

User-centred design focuses on the people using a product. It is about improving their wellbeing by creating a positive experience and making them feel comfortable and safe. It is about acknowledging the wide range of differences in humans from a physical, behavioural, lifestyle and cultural point of view. It is an approach to design that puts the user at the centre of all decisions and quite often involves ergonomic thinking. It might also include the addition of secondary **functions** to a product to make it more suitable for a specific user.

Activity 11.2a: Purpose, function and context and user-centred design

Give an example of how the products in the table below could be designed with different features or secondary functions (from the Purpose, function and context factor) depending on the user.

Product: Torch	Primary function – a portable light source
User	Features and secondary functions for the user's specific needs
A dentist looking into a patient's mouth	
A woman who works late and has a long walk up a dark path to her front door	
A person in a wheelchair who often needs to get out of bed and into the wheelchair at night unassisted	
A child who likes to hide under the bed and play with toys	

Product: Bag	Primary function – to carry things
User	Features and secondary functions for the user's specific needs
An athlete going to training	
Woman going to a formal event	
The household grocery shopper	
Travelling salesperson	
Product: Glasses case	**Primary function – to carry glasses**
A beach-goer with sunglasses	
A person who only needs reading glasses	
A child with two pairs of glasses for different distances	
Product: Umbrella	**Primary function – to provide shade or rain protection**
A beach-goer	
A female city worker who needs to carry the umbrella in her handbag	
A teenage umpire	
A couple who always go for walks together	

Activity 11.2b: User-centred design using lateral thinking and ergonomics

(For information on **lateral thinking**, see page 53–4, and for **ergonomics**, see pages 101–2 of your Nelson textbook.)

Consideration of user-centred factors may not be as simple as applying different features or secondary functions to a product. It may involve a whole rethink about how the product might be used and what would suit the user.

Make suggestions in the table on the following page for seating solutions that may not be a predictable response, or define a particular product (avoid the usual solutions).

| **Problem**: new solutions for somewhere to work, sit or lie ||
User	Suggested solution
A teenager with all afternoon to read a book	
An office worker who works at a computer all day	
Tradesmen at their morning break	
A child in a classroom in a developing country with dirt floors and no furniture	

Ergonomics

Ergonomics is an important aspect of user-centred design as it involves an understanding of the human body and how it is used. Choose one of the users from the above table and explain how the use of ergonomics or anthropometric data could assist a **designer** in creating a solution and what other specific measurements they would need.

_____

_____

_____

Activity 11.2c: User-centred design research methods

Bette is a carpenter who needs something to hold her 'tools of trade' that she can either attach to or wear on her body while working. Currently she carries a toolbox and places it on the floor, but this is not very convenient when she is working on a ladder. She finds that the available tool belts are uncomfortable and get in the way. She has approached a designer to solve her problems as she thinks many carpenters would purchase a product that provides a good solution.

Have a class discussion and from the pairs of research methods in each oval shape on the next page, highlight or circle one method in each oval that you think best suits a user-centred design approach.

Use another colour to highlight the methods that would require finding ergonomic information. Make a note beside it as to what sort of ergonomic information might be needed.

Observe her working
—
Look at photos of similar workers

Try out her role for the day
—
Try to imagine her typical daily activities

Watch her activities all week
—
Assume that she works mainly on building framing, cladding, doors and windows

Observe her arriving and leaving for work
—
Only take note of her activities when building

Adapt an existing belt design by an extra pouch
—
Take measurements of all the tools she carries around

Copy a design in a magazine
—
Make a prototype and ask her to use it and observe how it works for her

Adapt the design of a traveller's bag
—
Find out if she has seen anything similar to what she would like

Ask her questions and listen to the answers
—
Tell her that you know what she needs

11.3 PRESENTING WORK TO OTHERS

In Unit 2 you will need to present your work to others, whether they are your school collaborators or the end-user/s you are designing for. This is an important skill in any design career and will help to build your experience and confidence in justifying your ideas.

You might be asked to:

- describe the results of research you have carried out
- explain your group's preferred design choice to the rest of the class
- explain and demonstrate in 30–60 seconds what you and your group are up to
- demonstrate the features of the finished product/s at the end of the unit.

Methods that could be used are:

- speaking with slides or referring to an object
- sitting around and referring to drawings, samples or models
- creating a digital file with text or audio.

CHAPTER 12
Working as a team: Investigating and defining

Unit 2, Outcome 1
12.1 Approaches to teamwork .... p. 79
12.2 Team scenario .... p. 80
12.3 Developing the team design brief .... p. 82
12.4 Finalising the briefs .... p. 84
12.5 Developing evaluation criteria .... p. 84

12.1 APPROACHES TO TEAMWORK

There are various approaches that can be used to manage the teamwork and individual input in Unit 2. (For more information on team approaches, see pages 273–4 of your Nelson textbook.)
Teams can work on:
- products with a tightly defined focus (e.g. a range of containers or bags in a particular style for a specific **target market**)
- products with a loosely connected theme (e.g. a range of products made from bamboo)
- one product where team members make a different part. The product would need to be complex to allow for enough input from each student in the team (e.g. a stage set or costume outfit for a play).

There may be other ways that your class approaches teamwork or variations on the approaches above. Whatever your approach, it's important that responsibilities in every step are clearly defined, and that the team work together when necessary and agree on the influencing style or movement.

Team strengths and size

Analyse the strength of the individuals in your team. Consider the aspects of the design process that each of you loves doing as well as the things you each do well or competently. Cover all stages of the design process.

> **TIP**
>
> **Team size**
> Most teams work well in small groups. Aim for your team to have between three and five members. Any more than five team members, and the work of the team can become difficult to manage.

Organising your design folio

Your folio work this semester will be a combination of teamwork and your individual work. Teamwork should reduce your team's workload because you can share information and some tasks. Your team will all work together on some elements of your folio, then copy the end results so all members have a version in their folio (e.g. **design brief** and **evaluation criteria**). Some elements will be done individually and then shared with the team (research, planning components), and other elements will be your individual work (e.g. design and development tasks).

For assessment, you need to clearly identify:
- the parts of your folio that are collaborative team work or evidence of team feedback
- work done by others in the team
- your individual work.

If your team is working on one large product, you may be asked to produce a team folio that everyone in the team contributes to. Again, it is important to identify the work that is completed by the team, and the tasks contributed by individuals within the team. (It might be helpful to have a coversheet that lists the folio tasks and identifies who did what.)

Your folio should be structured to follow the main stages and steps in the product design process, although you don't have to do these steps in exact order. Use the stages and steps as headings for your folio work.

Your teacher will give you guidelines and an assessment framework that will clarify how they want you to organise your folio and the balance of team and individual work.

Activity 12.1: My team

As a team, discuss your individual design and production strengths and interests. This will help you to effectively plan your teamwork, and will allow your teacher to understand your choices in the allocation of responsibilities.

Record this information in the table below, or you can create your own. Add this to your folio.

Members	Strengths and/or preferences
e.g. 1: Chris Reed	Likes sketching and using Adobe Photoshop
e.g. 2: Ange Mason	Likes working with the fine details in metal

12.2 TEAM SCENARIO

A scenario is a brief outline of the design problem, opportunity or need for which your team will create their design brief.

You also need a statement that reflects the user-centred need (what is the main purpose of the product for users) – this will keep the group focused on the people who will use the final product.

Sample scenarios

Example 1

A shop that sells fishing gear wants a window display that reflects the stock available inside. They want a setting that suggests a fisherman's camp from the 1950s, such as a stool, boots, two hats, a carry bag and a box to display fishing lures.

User-centred need: creating products that are helpful for, and increase the enjoyment of, people who go fishing.

Example 2

A childcare centre needs a range of children's toys and/or games (either hard or soft) that encourage children to learn, suit different abilities and age levels, and can be easily stored when not in use.

User-centred need: developing toys/games that are safe, fun and educational for children of different ages and capabilities.

Example 3

You and your friends are all in the school play as angels (or witches). You are required to wear outfits that reflect that theme and have a similar colour range. The influencing style is early 20th century Disney films.

User-centred need is: creating clothing for a performance, which involves a lot of social interaction and gives satisfaction and enjoyment. Costumes need to be comfortable for the performers and allow for fast changes.

Example 4

Your team has been asked to design a sports uniform/outfit for a local basketball team in a style based on Melbourne street art. It needs to include a two-piece playing uniform (with a top and either a skirt or shorts), outer warm clothing and a carry bag.

User-centred need is: the basketball team members need to recognise other team players easily, and it has to fit them all comfortably to play basketball in as they are all very tall.

Example 5

The owners of a high-quality homewares shop in a country town want a range of innovative homeware products developed to highlight local materials with a modern style.

User-centred need is: the customers need useful homewares that function well and are visually appealing, but are also recognisably from the region.

Activity 12.2a: Brainstorming opportunities or areas of need

Set a timer for five minutes. Each team member has a blank page and writes three or more suggestions for your team's opportunities/needs, or based around a broad area of need given to you by your teacher. Next, use a brainstorming concept map (see page 200) to write down the best scenario ideas from the group.

Add both your page of suggestions and the best group ideas to your folio with clear headings.

Activity 12.2b: Summary of team scenario

Set the timer for five minutes again. As a team, select the most appropriate design opportunity or need. Write a scenario (brief statement) that describes this and decide on the influencing style or movement and what the specific user-centred need for this situation is. Create your own page or use the table provided below. This needs to be approved by your teacher before you go ahead. Each team member should have a copy and add it to their folio.

Team scenario	
Style or movement to be researched and to be influenced by	
User-centred need	
Teacher approval and comment	

12.3 DEVELOPING THE TEAM DESIGN BRIEF

Once the design scenario has been decided on (Activity 12.2), you need to develop it into a design brief that explains the context and includes constraints and considerations (and relevant product design factors). Remember that a design brief is not a description of a product. It outlines and gives details of a design **problem**, opportunity or need. Each team member contributes details for the design brief. The design brief provides more detail than the scenario – the details will need to be created or imagined, or you will need to research the situation in more detail. (For more information on writing a design brief, see pages 12–18 and 277–8 your Nelson textbook.)

Have a group meeting and decide on the details of the design brief. There are several alternative activities on the following pages to assist:
- Activity 12.3 – for individual contribution of ideas for the team design brief
- Activity 12.4 – use this activity to collate members' contributions. Each member will need a copy.

You will have different ways of writing your design brief, depending on your approach to the collaborative design task. Choose the most suitable.

My group is working on a project that is (circle):

Multiple copies of one product	A complex product made of separate parts (e.g. a stage outfit, set design, outdoor setting)	Different indvidual products that make up a range
One team design brief with a single list of constraints and considerations	One short design scenario/brief with a few constraints related to range/theme, and a more detailed individual design brief with more constraints and considerations	

Activity 12.3: My ideas for the team design brief

Use the following activity to note ideas for your group design brief – spend 10–20 minutes on this before sharing your ideas and making decisions as a team.

Outline of the context

My suggestions – information to help outline the design problem and context, and ideas on how to incorporate the chosen influencing style or movement, a description of the user/s and their needs.

_____

_____

Constraints and considerations

My suggestions are as follows.
Function (Define the functional aspects required and any features that would contribute to the success or value of the specific purpose of the products, e.g. size, fit, safety. If needed, explain how the setting/context might affect the function of the product.)

_____

_____

Aesthetics/appearance (Define the theme, specific style or unifying **design elements** and principles of design, e.g. shiny sections, thick straight lines, geometrical long thin shapes or block shapes. This relates to your research on a historical or cultural style or movement in Activity 13.3.)

_____

_____

Materials (Define what is needed from the materials, e.g. need to be rustproof, light, recyclable, cheap, strong or whatever suits the product or range. How many materials need to be used?)

_____

_____

Quality (Suggest the quality expected, e.g. accuracy of joins to 0.5 mm, squareness, visibility of welds or solder, fineness of sanding, straightness of seams, symmetry of the garment, type of finish, no loose threads, etc.)

_____

_____

Budget/material limits and **time** (Check your deadlines and material availability with your teacher.)

_____

_____

Sustainability issues (Consider sustainability issues – could be related to materials sourcing, waste during production, durability, disposal, etc. Sustainability strategies could be included, such as recyclable, reusable, etc.)

_____

_____

_____

What additional aspects does our team need to consider? This could overlap with any of the above.

_____

_____

_____

Add this to your folio.

Team and individual design brief

If each team member is making a different product for different users and situations, you will need to add an individual brief under the team's brief. What do I need to consider for my individual contribution? Does it have any specific needs, requirements and constraints that are not outlined in the team brief?

_____

_____

_____

Add this to your folio.

12.4 FINALISING THE BRIEFS

Have a quick team meeting to decide on the final version of your team design brief, and to decide who does what. Start by looking at each member's suggestions for each of the product design factors (function, aesthetics, material requirements, quality expected, budget and sustainability) and choose the most suitable ones. Omit or combine those that are similar. If your team is working on a range of products with a loose theme, your team design brief could be very short. However, your section of the design brief or your individual design brief can have the necessary details or requirements related to your expected contribution. Your teacher will clarify the expected balance or size of team/individual brief.

The constraint for aesthetics/appearance could be written as: 'Must follow the team style as defined in our style board' (covered later in this workbook).

Activity 12.4: Summarising team and individual design briefs

Once each member's contribution to the design brief has been considered, and decisions are agreed on, write the team's final design brief. Use Activity 12.3 to assist you. Make sure all team members have an accurate copy of the team section of the brief. Add the individual design brief if needed. Check that your individual section of the design brief doesn't just repeat the requirements for the team; things only need to be written once! Each member adds a copy to their folio.

12.5 DEVELOPING EVALUATION CRITERIA

(For more information on evaluation criteria, see pages 19–21 and 278–9 of your Nelson textbook.)

Evaluation criteria for the finished product are to be drawn up in a table with four columns, as shown below.

Evaluation criteria	Reason or importance	Strategies to achieve	Checking method for finished product
Write the evaluation criterion as a question.	Explain why this criterion is important to the design requirements.	Suggest research, trials or exploration of ways to achieve this	State a specific way you will check the product when it is finished to see how it fulfils this criterion. This information will help you to write the evaluation report at the end of the unit when your product is completed.

You will use the same criteria to judge the best design option but for this you will only need the evaluation criteria questions. Rewording or abbreviating them for this purpose may make it easier for you.

These questions need to be 'answerable' from a drawing and should cover: the appearance (i.e. influence from researched style or fits with team theme), the proposed materials, functional aspects, suitable proposed construction methods, sustainability and general suitability for the user.

In Unit 2, evaluation criteria for the product should cover these aspects:

- **functional** aspects to suit the user and **user-centred design** parameters such as: ergonomics and safety, how it satisfies the user's needs and wants, etc. (this should come directly from the design brief outline of context or constraints/considerations)
- **aesthetics** – the historical or cultural style or movement researched and its influence on the finished product or the unifying aspects of the style or appearance on the team's work
- suitability of **materials** used
- **quality** (the standard of quality expected by the team)
- **sustainability** issues related to the design, materials and processes involved in making the product
- **budget and finish date**.

Examples of evaluation criteria for the product

In the following table, two examples of evaluation criteria are given for the product design factor of purpose, function and context. Refer to the product design factors in your design brief and use them as a basis for your evaluation criteria questions.

	Evaluation criteria	Reason or importance	Strategies to achieve this	Checking method for the finished product
Function	Does the pot-moving device go through the gate easily?	There is a gate along the side of the house to access the backyard. The wheelbarrow must fit through to be useful.	I will measure the gate width and allow for 'wriggle' room.	When the wheelbarrow is finished, we will take it through the gate and check to see that it fits through easily.
	Are the angel costumes easy and quick to change in and out of?	We need to be able to change quickly to fulfil our multiple roles.	I will explore several ideas for fasteners that are secure but easy to undo.	When the costumes are finished, we will change in and out of them to see if it's easy and quick to do.

Examples of evaluation criteria for other factors

For a pot-moving device (like a wheelbarrow)

Product design factor	Example criteria questions
Aesthetics (visual, tactile)	Does the pot-moving device have a sleek, futuristic style as outlined in the design brief?
Style (could relate to both materials and visual)	Are the decorative elements and colours in keeping with the sleek, futuristic style chosen?
Material requirements	Are the materials used durable for outdoors?
Quality (could relate to materials, construction, etc.)	Is the device well made and finished so that it is water-resistant?
User-centred design	Is the device well balanced and easy for an elderly man to lift and manoeuvre?
Sustainability*	Are the materials sourced locally? Did I create a lot of waste during production? Is the product made to last a long time? Can the parts or materials be separated/disassembled for recycling?

(For more ideas about 'sustainability' evaluation criteria, refer to and adapt the social and environmental design strategies outlined in Chapter 5 (specifically pages 124–6) of your Nelson textbook.)

When creating your own individual evaluation criteria, cover what is relevant to your design situation, the user and the requirements of the materials.

Activity 12.5: Team and individual evaluation criteria

Discuss the evaluation criteria related to sustainability issues (marked*) with your team or class. Which issues are relevant to your product, which can you research effectively, and what areas can you control or make decisions about within your school setting? How will you judge how sustainable your finished product is?

All your team members should have identical evaluation criteria for the team's objectives and different ones for your own contributions. However, the team needs to collaborate and approve all evaluation criteria.

> **TIP**
> Make sure that in either your team or individual evaluation criteria, you cover the product design factors (particularly user-centred design), the style or movement that was researched, and sustainability.

Team evaluation criteria
(can be collated by one member of the team and distributed)

Evaluation criterion question	Justification – reason for importance	Strategies to achieve this	Checking method on the finished product

My individual evaluation criteria

Evaluation criterion question	Justification – reason for importance	Strategies to achieve this	Checking method on the finished product

Add extra rows for each criterion as needed and add these tables to your folio when complete.

CHAPTER 13

Research

Unit 2, Outcome 1

13.1 Primary and secondary sources of research	p. 87
13.2 Planning for research	p. 88
13.3 Researching a historical or cultural style or movement	p. 89
13.4 Process trials and materials testing	p. 92

13.1 PRIMARY AND SECONDARY SOURCES OF RESEARCH

(For information on primary and secondary sources of research, see pages 25 and 279–80 of your Nelson textbook.)

Activity 13.1: Sources of research (suitable for all units)

Primary research is data, images and information that you have discovered or collected yourself. Secondary research is done by other people, and you can find their information on the internet, in a book, on TV, etc. The research methods listed below come from either primary or secondary sources. Choose different-coloured highlighters for primary and secondary sources and colour the methods accordingly. Indicate your choice of colours in the two boxes below (the key).

Primary source [ ]

Secondary source [ ]

Taking photos of people and their clothing	Copying photos from an Internet site	Cutting photos from a brochure or magazine	Sketching from direct observation
Getting average reach measurements from an ergonomics book	Finding out current trends by checking out a number of fashion/interior design blogs	Getting text from the Internet on the characteristics and properties of silk	Getting the details of the Australian Standard for a product from the Internet
Practising several methods for joining	Measuring a user's space or body	Burning material samples to check their flammability	Going into shops and looking at similar products
Looking up joining methods in a textbook or magazine	Washing samples to check for reaction	Going to an event to observe the behaviour of participants	Testing and measuring the flex between radiata pine and oak

Think about the difference between primary and secondary sources of information when you are doing your research, and make sure that you use both.

13.2 PLANNING FOR RESEARCH

Activity 13.2a: Research plan

Sit with your team members, set a 5-minute time limit and list as many areas as possible on which your team will need information and ideas. Remember that in Unit 2 you are to use the user-centred design approach, which requires you to give a lot of thought towards the user of any product.

A mind map template can be found in the Design Folio Template on page 200. Write the design focus in the centre of the mind map. From your design brief, note areas that are important requirements, related to:
- the users and their specific needs
- their situation (where the product will be used)
- the purpose/function of the product
- possible styles
- any other relevant product design factors, such as size, cost and materials.

This will help direct your research. Use this to start thinking about how you will share the research tasks across the team. Add this to your folio.

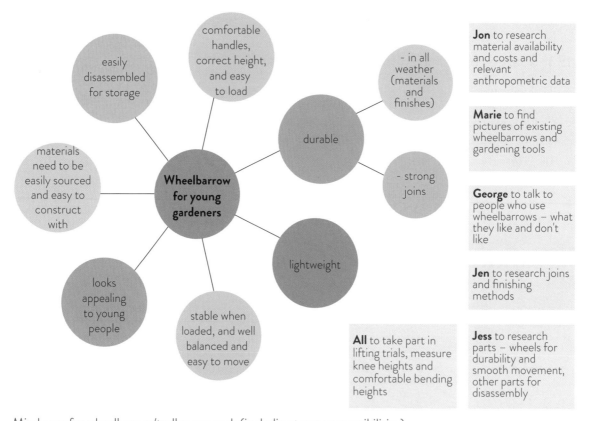

Mind map for wheelbarrow/trolley research (including team responsibilities)

Activity 13.2b: Research plan

Use the table provided on the next page or a different graphic organiser to allocate research responsibilities. Include some methods of research or research activities that will assist you and your team to best meet the specific needs of the user/s of your product.

Add this to your folio.

Areas for research	What sort of information is required	Who is responsible and how it will be sourced
Purpose, function, context: • Primary function • Possible secondary functions • Where it will be used • Existing products		
Materials: • Characteristics and properties • Trials and testing • Costing		
Joins and other **processes**		
Measurements and **sizes**		
User-centred design (list areas to cover)		
Historical/cultural **styles** (list areas to cover)		
Other visual aspects – colours, features, existing ideas		
Sustainability (list areas to cover)		
Other areas		

Highlight the areas of research you are responsible for.

Use the following pages and/or the research pages in the Design Folio Template to plan and present this research. Complete further research specifically needed for your component or individual product.

13.3 RESEARCHING A HISTORICAL OR CULTURAL STYLE OR MOVEMENT

(For information and images on various styles and movements, see Chapter 11 of your Nelson textbook. You will need to do further research and gather several images of your chosen style.)

The following activity can be completed before or after the design brief is written to determine a theme (and its stylistic elements). This would be research from a secondary source.

Activity 13.3: Research on a style or movement

After looking through Chapter 11, 'Design inspiration', and discussing with your team, select a historical or cultural style or movement. This could include those in the table on the following page.

A historical style or movement	A fashion period	A cultural style
• Art Deco • Bauhaus • Memphis • Mid-Century Modern	• 1920s jazz • World War II (1939–45) • 1950s glamour • 1960s Pop and Op Art • 1990s grunge	• Gothic • Oriental • Steampunk • Organic • Minimalist

1. Find at least three related images of products and insert or paste them in a table such as the one provided below and on the following page, or create your own page layout.
2. From your research, describe the function, materials, quality and the typical use of design elements and principles in each product (i.e. how the style uses the design elements – line, shape, form, texture and colour – and the design principles, such as balance, contrast, how the proportions are divided, use of mass and space etc.).
3. Pinpoint four or five main features that are commonly seen in the products and furniture of the style or period (if relevant).
4. Include any relevant underpinning philosophy or approach and/or provide some of the historical and social reasons or causes for the design decisions made.
5. If possible, find out the significant designers within this style.
6. Describe your response to this style/movement:
    - What do you like about this style/movement?
    - How could these aspects inspire you when creating a team or individual product?

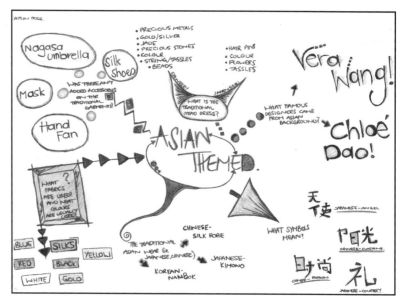

Concept map by student Caitlin Hogg identifies design directions from research

Name of style or movement: Approximate time or period (if relevant):	
Image 1 (Annotate – use of design elements and principles, use of materials, joins and processes)	**Background** of style or movement (social, historical, cultural)
Caption for above image (designer, name of piece, year):	

Image 2 (Annotate – use of design elements and principles, use of materials, joins and processes) Caption for above image (designer, name of piece, year):	**Common design features** (use of the design elements and principles)
Image 3 (Annotate – use of design elements and principles, use of materials, joins and processes) Caption for above image (designer, name of piece, year):	**Significant designers**

What do you like about this style/movement?

How could these design features **inspire your designs**?

13.4 PROCESS TRIALS AND MATERIALS TESTING

Process trials and material testing can be carried out at different times during the product design process. It might be important to perform them before the design brief, straight after it, after the preferred option has been selected or during production. This would be research from a primary source – your own experience. (For information and suggestions on material tests, see pages 151–7 of your Nelson textbook.)

For instructions about how to carry out and report on a materials tests, you can refer to Activity 2.6a in the 'Design fundamentals' section. You also need to carry out some process trials to see which methods are the most suitable (use Activity 2.6b) and what skills you might need to practise to achieve quality. Use the following activity to plan and record your results.

Activity 13.4a: Process trials

The processes being practised or compared are:

1 _____

2 _____

Equipment needed:

_____

_____

For _____ (product or component), we need the following standard of quality in the process (think about the purpose of the process and the standard required for it to work well):

_____

_____

To trial this, we are going to (e.g. 'compare these two processes'):

_____

_____

We will judge the results by:

_____

_____

Trial results

Process 1 (insert a photo or diagram)	Results and comments

Process 2	Results and comments

Decisions from the trials

The most suitable process for _____ is _____ because:

_____

_____

Add this to your folio.

Activity 13.4b: Sustainability research on selected materials

In Unit 2, you need to make a judgement on the materials you are using in terms of their sustainability. This will require research from a secondary source. Select at least two materials that could be used for your product.

Product you are making: _____

	Material 1	Material 2
Name of material:		
What is the raw state of these two materials in nature?		
Where are the materials sourced and processed (distance travelled)?		
Explain how each material is suitable for the situation and how it makes the product sustainable.		
Can you reduce the total amount of material used because of the strength of the material (relevant for WMP materials)?		
How easy is this material to recycle to re-use or to decompose?		

Extend the table if you are researching more than two materials. Add any other sustainability issues relevant to these materials. Add samples of the materials if possible. If not, then add images. (For information on each material, see Chapters 7 and 8 of your Nelson textbook. Complete 'Design fundamentals' Activity 2.7: Materials and sustainability.)

Add this to your folio.

Sustainability issues related to your intended product

On a new page or in the space provided, complete research and respond to as many of the following dot points as you can.
- How could your intended product benefit people (social sustainability)?
- If your product went into mass-production, what social issues might it cause or benefits could it bring to others, i.e. factory workers, manufacturers, retailers, user/s?
- List some negative issues that the mass-production of your product could create.
- What are the possible negative sustainability issues surrounding use of your product, i.e. cleaning, maintenance, ease of replacement of parts, energy required for use, life span?
- What can be done with the materials when the product is no longer useful? Can they be reused or refashioned?
- What methods could you use to reduce the use of materials (avoid waste) during the 'cutting out' step?

_____
_____
_____
_____
_____
_____
_____
_____
_____
_____
_____

Activity 13.4c: Research results

Use the research information page of the Design Folio Template (page 179) to plan and complete one or more pages of research about other areas you have identified. Add these to your folio or to share with others within your team. Try to represent your research in the most visual manner possible. An example is shown here at right.

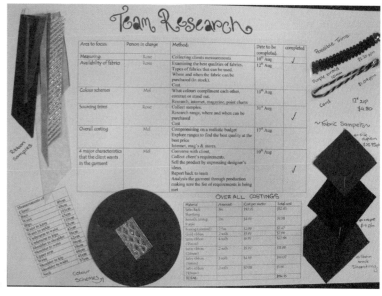

An example of presenting the team's research, by students Rose Dwyer and Melinda Stephens

CHAPTER 14

Designing within a team: Design and development

Unit 2, Outcome 1
- 14.1 Designing within a team .... p. 95
- 14.2 Creating a style .... p. 96
- 14.3 Visualisations and designing creatively .... p. 97
- 14.4 Developing your design options .... p. 99
- 14.5 Selecting and justifying your preferred option .... p. 100
- 14.6 Working drawings .... p. 102

For students who are new to the study, many activities from the 'Design fundamentals' section are suitable as an introduction for the design and development stage of the product design process. They are:
- Activity 3.4a: Using the design elements and principles
- Activity 3.4b: Inspiration from elsewhere
- Activity 3.4c: Trigger words
- Activities in 3.5 and 3.6: Drawing information, techniques and activities

14.1 DESIGNING WITHIN A TEAM

(For information on the definitions of visualisations and presentation drawings and what is expected in the design and development stage of the design process, see pages 58–71 of your Nelson textbook and 3.1–3.3 in the 'Design fundamentals' section.)

The activities in this section may be completed in the order that suits your team. You may prefer to create design ideas first and then define the team 'style' or you might prefer to define the team's 'style' beforehand.

Activity 14.1: Design responsibilities

Complete the design responsibilities table below. Remember to work with, and use, the skills of your team. Add this to your folio. Any time waiting for team members could be well spent practising production processes.

Stage	Responsibility
Brainstorming, creative thinking and mood board (to redefine style and how design elements will be used)	Whole group
Visualisations – idea and concept sketches	
Feedback on idea sketches	Whole group
Design options as presentation drawings (follow teacher's instructions on number required)	
Feedback on design options and selection of best design(s)	Whole group
Working drawings (define type here, such as flats, pattern pieces and the construction diagrams needed)	

14.2 CREATING A STYLE

Styles are created by the combination of design elements (what is used) and the design principles (how these elements are used or arranged). They may also have an underlying philosophy. The style or movement that you researched in Activity 13.3 will assist you here.

To create a style for your project, your team needs to define how the design elements and principles will be used consistently so that the finished product has a unified and recognisable style. Decisions need to be made and documented regarding the type of line to be used, the shapes, textures, colours and forms and how each member's contribution could complement or match. Then further decisions need to be made on how these will be arranged (i.e. how the design principles will be used). Some aspects to consider are: Symmetrical or asymmetrical? Striped thick, thin or irregular lines? Colour scheme? Material colours? Patterned or plain? Regular or unusual proportions? Decoration? Rounded shapes or hard edges?

Mood board or team style guide

(For information on mood boards, see pages 56–7 of your Nelson textbook.)

A mood board or inspiration board is a collage that can be used to depict a 'mood' or to set specific features of a style. A mood board can be made up of drawings, paint or coloured paper, photos, pictures cut out of magazines (e.g. possible shapes and colours to reflect the theme), textures such as fabrics or photographs of surfaces and trimmings (actual pieces or photos), and photos of places where the design will be used (context). A mood board may also be created digitally using a collage app.

In Unit 2, a mood board can be used in any manner that you like to assist your team in defining common elements. To see examples of mood boards, type the term into a search engine on the Internet. But beware! Many pieces called 'mood boards' are just a mish-mash collage without depicting any 'mood' at all!

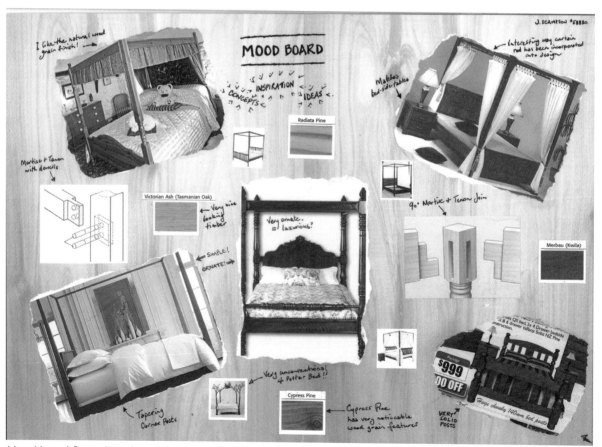

Mood board for a client by student Jonathon Scampton

Activity 14.2 Mood board

Refer to page 181 of the Design Folio Template and create a mood board by:
- pinning images, objects, photos of components, thin material samples, examples of finishes, etc. onto foam or cardboard, then taking a digital photo, saving it and printing it out for each team member
- digital methods, using a program or app for visual and text elements.
Add this to your folio.

14.3 VISUALISATIONS AND DESIGNING CREATIVELY

Activity 14.3a: Visualisations (idea formation)

Review the information on visualisations in Chapter 3, pages 29, 59–61 and 69 of your Nelson textbook, and use page 183 of the Design Folio Template to complete 1–3 pages of idea sketches showing the influence of your research, with the heading 'Visualisations'. Create short, quick sketches or 3D models that help to explore your ideas. You will be showing these ideas to your team for feedback, so you need to adhere to the team schedule. A good idea is to set a specific time (e.g. 20 minutes). Use pencil, black fineliner and colour so that your drawings have a strong impact. Annotate your visualisations and comment on:
- what works and what doesn't work in terms of function and satisfying the user's needs
- use of materials
- construction suggestions
- what works or doesn't work visually
- how your ideas match the team's style guidelines to reflect the chosen style or movement.

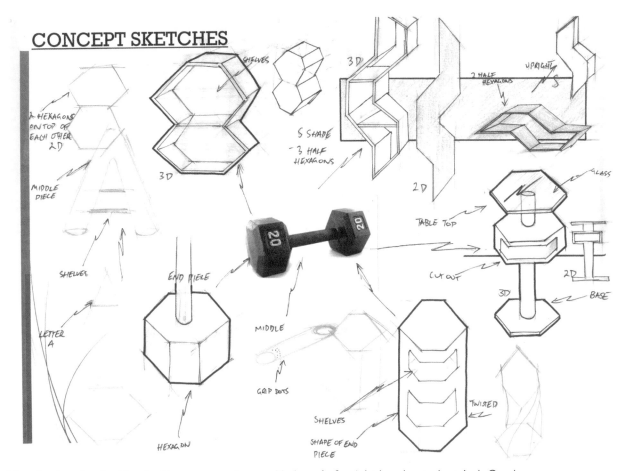

Visualisations inspired by the 'form, proportions and balance' of weight bars by student Josh Goudge

Activity 14.3b: Designing with style

Look at your visualisations and start to develop some of them further in terms of their aesthetics or appearance (as you did in Unit 1) with reference to your team's style guide. Do this by changing proportions, changing the shapes (by adding curves, repeating shapes, separating portions with a line, or blending components into a whole, copying shapes in nature and applying patterns). Re-explore functional aspects.

Set a time limit (e.g. 20 minutes). Create 1–2 more pages of these and add them to your folio. Use colour and bold outlines.

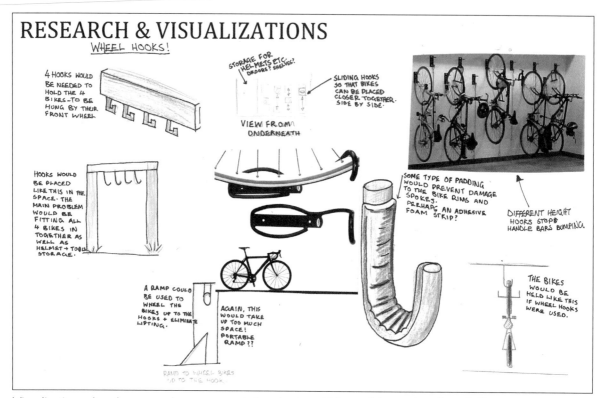

Visualisations placed among relevant research for a backyard bike rack by student Sam Jeffrey

Activity 14.3c: Group feedback from brainstorming and visualisations

After a set time, get together with your team mates and discuss each other's ideas with reference to the team style guide or mood board. Give each person 1–2 minutes to quickly explain their ideas, then provide constructive feedback.

Record your team mates' comments on your design ideas (you also need to comment on their ideas).

Create a table and write down at least one comment from each team member. An example and table are provided below. Add this to your folio.

Ideas and sketches I've completed	
Two A4 pages of thumbnails that explored different shapes for the wheelbarrow tray and where to place the wheels.	
Feedback from:	Strengths/Weaknesses
Ange	Couldn't see enough detail, but really like some of the shapes that Chris has used in the first three sketches.
Lee	Didn't think any of the joining methods would be strong enough, but we could develop the idea in drawing number 7.

Feedback on my ideas

Ideas and sketches I've completed	
Feedback from:	Strengths/Weaknesses
	_____ _____ _____
	_____ _____ _____ _____
	_____ _____ _____
Best ideas in the following areas: • Function • Appearance and aesthetics • Material use • Sustainability	

Add or delete rows as needed.

14.4 DEVELOPING YOUR DESIGN OPTIONS

(For information on design options and the presentation drawing style, review the relevant section of Chapter 3, and see pages 30–1, 61–3 and 69–70 of your Nelson textbook.)

Activity 14.4: Presentation drawings for design options

Design options are presentation drawings of the best ideas from your visualisations. They should also be viable solutions for the team design brief.

Design options need to be annotated to show:
- hidden features or hidden construction methods
- possible or definite materials to be used
- how different features or aspects of the design relate to the design brief
- how the team's 'style' has been incorporated.

Use the design option sheet of the Design Folio Template (page 184) and create at least two design options each (check with your teacher how many are required). Add them to your folio.

14.5 SELECTING AND JUSTIFYING YOUR PREFERRED OPTION

(For information on using evaluation criteria to select the preferred option, see pages 19–21 and 278–9 of your Nelson textbook.)

In all VCE units you are expected to use evaluation criteria to assist you in selecting the preferred option. This is one technique to assist your critical thinking as it requires you to make decisions. You are also expected to use feedback when selecting the preferred option. It is essential in Unit 2 to gather feedback from your team members.

Activity 14.5a: Group feedback on design options

When giving feedback to team members on design options, along with how the designs fulfill the design brief requirements, consider the following:
- Does the option have enough detail for judgement to be made?
- Does it fit with the team's goals and requirements?
- Can it be made with the existing skills, time, space, materials and equipment available?
- Will it fit the budget?
- Is it of a reasonable degree of simplicity or complexity, suitable for VCE, and can it be made in the time allowed?

Create your own table, similar to the one following, and write comments from the other team members for each of your design options according to each of your evaluation criteria. Add rows or columns as needed and add this to your folio. (For information on critical thinking, see pages 53–5 of your Nelson textbook.)

Feedback from:	Design option 1	Design option 2

Activity 14.5b: Selecting the preferred option using a criteria grid (optional)

When you apply evaluation criteria to each option, rate them by giving a score out of five. The score will be based on your own judgement and feedback from your group.

Draw up a table, or use the one following, for yourself and each team member to record their feedback. Write your evaluation criteria in the first column, a score for each one in the middle columns and a comment on which option/s best satisfies the criterion in the last column. Add this to your folio.

Selecting the preferred option

| Evaluation criteria | Design options | | | Comments/reasons for ratings |
	1	2	3	
1:	/5	/5	/5	
2:	/5	/5	/5	
3:	/5	/5	/5	
4:	/5	/5	/5	
5:	/5	/5	/5	
Totals	/25	/25	/25	

This method can be used for the design options of each team member.

Activity 14.5c: Justifying the selected or preferred option

The choice of preferred design option needs to be explained in a written justification. This includes team feedback and/or the scores for each option. Regardless of your team's approach, all team members should have input into the choice and be in agreement. Two examples are given.

> **Example 1: For a team product**
>
> Our group chose Design Option 2 by Chris as the design to make into a wheelbarrow. We all liked it because it looks different than a normal wheelbarrow, almost a cross between a trolley and a wheelbarrow. We all liked the tip truck type features and thought that it would be easy for Mr Smith to load his pots on and off. We also thought it had beautiful shapes and patterns on the side that Mr Smith would like. However, we are going to include the designs for the handles from Jay's design option 1 as they are long and sleek looking and we know we can make them from the metal that we have. We still need to decide on the way to join the wheels to the front section.

> **Example 2: For a range of products**
>
> After getting feedback from my team and using a criteria grid to compare the scores, it was decided that I should make Design Option 2 for the soft toy. It scored the highest with 22/25 and it will be easier to include all the 'style' elements of our group, and it looks like I will be able to finish it in the time frame, whereas Design Option 1 is more complicated and would need more time.

Justification for preferred option

The design option that was selected and/or scored the highest was design option number: _____.
Drawn by (optional): _____

Justification with reference to the evaluation criteria: (briefly explain its score and how the selected [preferred] option satisfied each evaluation criterion)

_____
_____
_____
_____
_____

Group feedback: _____
_____
_____
_____

Add this to your folio.

14.6 WORKING DRAWINGS

Working drawings need to be drawn (or created with CAD) for your preferred option to show the **product specifications** (i.e. all the information needed to construct it). They are usually 2D and are often drawn using the **orthogonal drawing** method for wood, metal and plastic products, or as '**flats**' for textiles products. 3D drawings can be included to aid clarity. Include accurate measurements, and you also need to draw accurate construction details – either on that drawing, or drawn and described separately.

Working drawings are important so that mistakes that waste material or time are avoided and to work out the details of construction. If you are working from a published design or pattern, you need to isolate the sections that apply to the product you are making. Adjust them as necessary. Enlarged drawings of details can assist in depicting small areas.

Activity 14.6: Detailed working drawings

Add to your folio either:
- a detailed working drawing of the team's chosen option (if you are working on a collaborative project) completed by one team member and shared
  or
- a detailed drawing of your preferred option (if the team is developing a range of products).

Complete the necessary working drawings (3D isometric, 2D orthogonal, 'flat' drawings, pattern pieces or shape templates, construction diagrams) using page 186 of the Design Folio Template and add these to your folio. (For information about working and detail drawings, see pages 33–5, 63–6 and 71–3 of your Nelson textbook. You can also look back at the instructions in Activity 3.7 in the 'Design fundamentals' section of this workbook.)

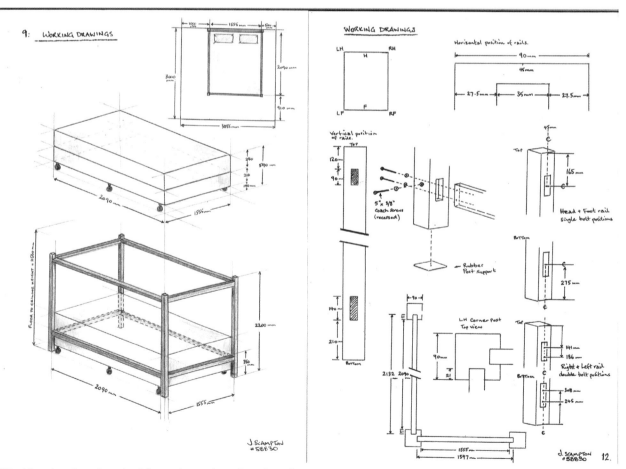

Working drawings for a bed frame by student Jonathon Scampton

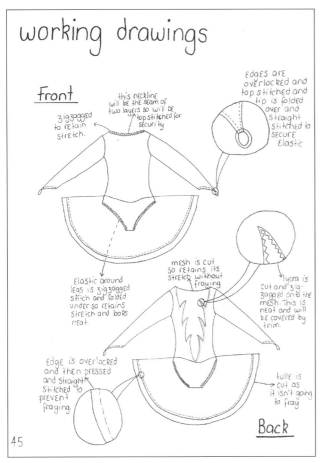
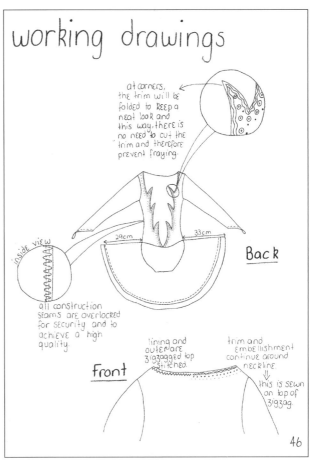

Flat drawings for a ballet dress by student Sian Banfield

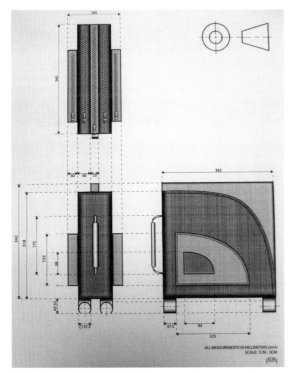

Orthogonal drawing for a clip-on bike pannier by student Steven Lam

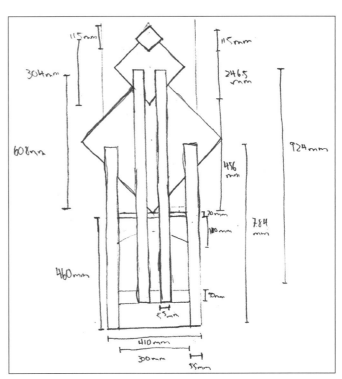

Hand-drawn orthogonal drawing for a chair by student James Fay

CHAPTER 15

Planning and production

Unit 2, Outcome 1
15.1 Scheduled production plan .... p. 104
15.2 Production steps .... p. 105
15.3 Timeline .... p. 105
15.4 Materials and costing list .... p. 106
15.5 Safety and risk assessment .... p. 106
15.6 Quality .... p. 106

15.1 SCHEDULED PRODUCTION PLAN

(For information on all the planning components required, see pages 36–43 of your Nelson textbook.)

Activity 15.1: Planning responsibilities

Complete the planning responsibilities table below. You will need to work out which items require both a team version and individual versions.

If you are making individual products, you can share research with the rest of the group for the materials tests and process trials, but you will need to complete an individual sequence of production steps.

If your group is making parts of one product or multiples of the same design (in a production line), you need to coordinate the production steps carefully so that time isn't wasted, and members aren't waiting for their production stages.

Each team member should create their own risk assessment table, but it can include information from others in their team.

Add this to your folio.

Planning items needed	Person(s) responsible	Date to be finished by
Materials test and process trials (list them) (unless already completed in your research)		
Production steps (for each product if needed)		
Timeline		
Materials list and costing		
Risk assessment	One for each member	
Quality measures list		

Add copies of each planning document to your folio.

15.2 PRODUCTION STEPS

The production steps explain in detail the steps you need to follow to make your product. It's important to think carefully, before you start, about the ways in which you could logically order your construction. However, don't worry if your plans change during production. Consider the major steps needed through these stages:
- measuring and marking out
- cutting, shaping or forming
- joining or assembling
- decorating or embellishing
- finishing.

The production steps should include: a brief description of how the step will be completed, estimated time for each step, the materials and equipment needed, and the safety precautions for each step (note that you will be completing a more detailed risk assessment later – the safety column could be left until this is completed).

If you are making a component of a team product, it is important to discuss and coordinate with your team when planning, so that your component stages are completed at the right time.

Activity 15.2: Production steps

Use the suggested table layout for production steps on page 188 of the Design Folio Template to explain your predicted steps of construction. If you are working on one product as a team, add an extra column for noting each member's responsibilities.

15.3 TIMELINE

Create a timeline in a table to show the schedule of your team project (i.e. expected dates of the start, the finish and various stages). Check with your team members so that your dates correlate. If your team is working on one product, add their names to the responsibilities on your chart. A Gantt chart (named after U.S. engineer Henry Gantt, who first devised it) can also be used. It shows a detailed schedule for a project with many workers involved (i.e. expected dates of the start, the finish and various stages).

Example of a timeline

Steps	2 August	5 August	9 August	11 August	16 August
1 Measuring and marking	Jen and Bill				
2 Cutting pieces	Chris	C, J and B			
3 Joining			C, J and B	C, J and B	
4 ...					C, J and B

Activity 15.3: Timeline

Use the suggested layout on page 189 of the Design Folio Template to complete a timeline for your production. Be sure to include dates and name all the production steps.

15.4 MATERIALS AND COSTING LIST

Activity 15.4: Materials and costing list

Complete this as a team or individually. Check that all requirements are listed. Use the suggested layout of the materials and costing list on page 187 of the Design Folio Template and refer to your working drawing to help you calculate all the materials and components you require to make your product.

15.5 SAFETY AND RISK ASSESSMENT

Activity 15.5: Safety and risk assessment

Being safety conscious requires you to think ahead and be prepared. Think about the processes you will be carrying out, the possible risks and the safety precautions required to minimise these risks. Rather than repeat these precautions for every construction step, group the activities that require the same precautions together. Don't forget to consider the hazards involved in moving materials and your workplace environment (e.g. height of benches, dust, noise, etc.). Refer to the risk assessment you completed in Unit 1 to assist you.

This can be completed as a team task – split up and allocate the responsibility for the main processes, then compile your work to create a complete risk assessment.

Use pages 190–1 of the Design Folio Template to complete a risk assessment sheet for your production.

Example risk assessment

Processes that have safety issues	Risk or hazard (the thing that causes an accident)	Possible injury or damage	Risk likelihood/ seriousness (rate 1–5)	Precautions to prevent risk or hazard
Moving around the room in between processes	Obstacles that cause slips, trips or falls	Various injuries to limbs, back or head	3/5 and 2/5	Always keep floor space clear and pick up bits and pieces, keep cords neat and tidy and away from walkways

15.6 QUALITY

Activity 15.6: Quality measures

Use page 189 of the Design Folio Template to describe quality measures for your team product/s. Or you can add a quality column to your sequence of production steps.

For wood, metal or plastics, include: the expected quality of the material selected; accuracy of measuring and marking out, cutting out pieces, drilling, routing, joining pieces, assembly; and any tools or equipment needed or setting-up procedures for quality results.

For textiles, include: the expected quality of the material selected; accuracy of pattern placement and grain direction, cutting out pattern pieces, the seams (straight, no missed bits, bunching of thread, etc.), stitching and finishing (matching colour of thread); any symmetry required in pockets, collars, sleeves, etc.

Collect all these planning sheets (Activities 15.1–15.6) together to form your scheduled production plan – make sure all team members have copies in their folios.

CHAPTER 16

Production

Unit 2, Outcome 2
16.1 Production responsibilities..p. 107
16.2 Documentation – journal...p. 107

16.1 PRODUCTION RESPONSIBILITIES

Activity 16.1: Production responsibilities

Refer to your production steps chart frequently during production, particularly at the start of each working session. If you are working on a group product, your team should have allocated responsibilities on that chart. It might help to use a copy of the chart as a progress checklist to keep all the team members on task. Add this to your folio at the end of production to support your journal. Apply risk management at all times and refer to your quality measures to achieve the best possible quality in your work.

16.2 DOCUMENTATION – JOURNAL

You are responsible for documenting the production work you complete.
It's important to note:
- the date and the time taken
- what you did to complete each step
- the equipment, tools, machines and safety procedures that you used
- any change in the order of the sequence, responsibilities, planned processes
- problems, mistakes or errors that forced a change in plans
- time spent waiting or depending on team members, any assistance accepted or given by team members
- skills you have learnt or improved.
Use digital photos and captions to illustrate your journal information.

> **TIP**
>
> 'Modify' means to change, adapt, improve or alter. A modification means any change, alteration, improvement or adaptation usually carried out to better suit a particular situation. In terms of your design and production, it refers to anything that didn't follow the plans (the design, the sequence, the materials, the components to be used, processes and the equipment). You are expected to document any modifications.

Activity 16.2: Journal

Make sure you write your journal regularly – at least once a week. Use the journal format in the Design Folio Template (page 195) or create one that suits you. You will need a number of sheets. Don't forget to take photos of your working stages to insert into the document. Compile all your journal entries and add them to your folio after all your planning items.

CHAPTER 17 Evaluation

Unit 2, Outcome 2	
17.1 Evaluating the product	p. 108

Evaluation is important throughout the design process, but particularly at the end. Evaluation helps us to look at what was achieved, how it was achieved, and identify where the product and the process can be improved. Evaluation serves us for the future – we can learn from what we have done and make better products more efficiently.

In Unit 2, you are expected to use your team and individual evaluation criteria to evaluate your product and the materials used, including:
- user-centred design
- sustainability
- the historic or cultural style or movement researched and the influence it had on your designs.

You also need to make suggestions for improvement.

It is also useful to reflect on your teamwork – the collaborative methods used to design and make the product – and make suggestions for improvement.

The headings outlined in the next few pages will assist you with your evaluation in Unit 2.

Responsibilities for evaluation	
Using checking suggestions from evaluation criteria	Whole group
Feedback from each team member	Whole group
Writing an evaluation report	Each team member writes their own report

17.1 EVALUATING THE PRODUCT

Responding to your evaluation criteria

Before you can respond to your evaluation criteria, you need to carry out the method of checking that you suggested earlier. It's important that you do this with your team members.

Write down your evaluation criteria. Directly underneath (or in a table), write how you checked the product, and then use your observations and the group's feedback to comment on how your product has satisfied or fulfilled this requirement. Make suggestions for improvement.

Examples

Evaluation criterion	How it was checked	Observations, feedback and response	Suggestions for improvement
Does the wheelbarrow go through the gate easily?	When the wheelbarrow was finished, we filled it up with pots and each team member had a turn at wheeling it in and out of the gateway.	Jenny found taking the wheelbarrow through the gate a bit difficult. She said she felt like she had to really concentrate so that she didn't bump the pots off. Chris thought that perhaps the wheelbarrow needed to be 4 cm narrower to make this task easier.	An obvious thing would have been to make the wheelbarrow narrower, taking into account the true amount needed to manoeuvre it.
Does the soft toy match the summer range of clothes for the window display?	When the range was finished, we put it all together and discussed how well it matched.	Sam thought that the toy was a bit too furry and wintry looking to match the other products, but Jess and Lee said that the little jacket matched the fabric of the other products, so they thought it was quite appropriate.	Next time I could choose a fabric for the toy that is less furry and more suited to summer.

Historical or cultural style

Use the evaluation criterion that you wrote earlier in the unit for the historical or cultural style you researched.

Respond to it by briefly describing elements of the historical or cultural style or movement that can be seen in your finished product. Include whether you think this was a successful input to your team's work.

Sustainability factors related to your product and materials

Use the evaluation criterion that you wrote earlier in the unit for sustainability.

Respond to it by referring to your research into the materials used (Activity 13.4b) and considering the following questions:
- What effects could this product have on the environment? How do the materials used affect the environment when they are sourced and processed, and when the product is manufactured and disposed of?
- Where is the material sourced? Is it plentiful, renewable/non-renewable? Who processes it? What effects does this have on their health or their livelihood? Are they paid and treated fairly? How does this material compare with alternatives?
- Is this a good quality product that will be cherished and last a long time?
- Suggest further improvements that could reduce the negative impacts of this product on the environment.

User-centred design approach

Use the evaluation criterion that you wrote earlier in the unit for user-centred design.

Respond to it by considering these questions:
- How could your product satisfy the specific needs and wants of users and improve how they live?
- Explain how your team focused on the needs of the people who might use this product, particularly when doing research. Did the team consider the needs of one user or many?
- How could it be designed and made to better suit a specific range of users, e.g. poorer people, people with diminished ability or people with different cultural values. What secondary functions could be added (or removed) to make it more suitable for a wider range of users?

Activity 17.1: Evaluating your finished product

Complete your report (either handwritten or typed) by using the table below to assist you.

Evaluation criterion	How it was checked	Feedback and response
Evaluation criteria: ● ● ● ● ●		
Style: ● ●		
Sustainability: ● ●		
User-centred design: ● ●		
General evaluation comments about the product		
Suggested areas for improvement for the product in each of the above areas		
Name processes whose quality you (and your team) could improve in your production work. Explain how you could do so.		
How well did your team work together?		
Suggestions to improve teamwork		

Optional evaluation activity

To reflect on your practices and consider improvements for Unit 3, complete Activity 10.3: Evaluating your design, planning and production process, on page 71.

ASSESSMENT FOR UNIT 2

Name: _____ My team members: _____

_____

Our project/scenario: _____

My contribution: _____

_____

The shaded area represents items that should be in your folio.

Checklist	Self-assessment	Teacher's assessment A–E
TEAMWORK Team design brief and evaluation criteria Team research and planning		
Team production and finished product		
Team collaboration and project management		
INDIVIDUAL CONTRIBUTION (highlight work done) My contribution to:		
OUTCOME 1 **Design brief/Evaluation criteria**		
Research findings/Research presentation/Materials test report/Process trials/Information on material and processes being used Historical or cultural style or movement		
Brainstorming/Visualisations/Numbered and annotated design options/Use of feedback, evaluation criteria, decision grid and justification in selecting preferred option		
Working drawing/Timeline/Production steps/Materials and costing list/Risk assessment/Quality measures		
OUTCOME 2 **Production** • Contributed to the team's project (i.e. used time effectively and made an effort to follow team plans) • Carried out processes safely and accurately		
Documentation • Recorded information about work completed in a log or journal		
Finished product • Suitability of product as a solution to the brief • Quality and accuracy of workmanship • My contribution		
Evaluation report • Evaluation of the product with response to evaluation criteria (with reference to sustainability issues, historical or cultural style and user-centred design factors) and suggestions for improvement; reflection on teamwork		

UNIT 3

APPLYING THE PRODUCT DESIGN PROCESS

This chapter will cover the following areas in detail:

Area of Study 1: Designing for end-user/s

18 Designer's role, the process, factors and market research   p. 114

Area of Study 2: Product development in industry

19 R&D, technologies, sustainability, planned obsolescence and manufacturing   p. 127

Area of Study 3: Designing for others

20 Activities to prepare for the SAT   p. 144

UNIT 3 CALENDAR

Suggested student timetable – check with your teacher for the dates and highlight the activities you need to complete. Use a different highlighter to mark the work associated with the SAT.

Week	Topic	Work for assessment/submission	Due date
1	**Outcome 1:** The product design process, role of a designer, market research and product design factors	Activities on pages 114–20	
2	**Outcome 1:** Structure of a design brief, evaluation criteria; practise writing design briefs (not for SAT) and directions for research **Outcome 3:** Preparing for SAT – researching and deciding on end-user	Activities on pages 121–4 SAT folio – brainstorming, end-user research, interview and end-user profile DFT on pages 173–4	
3	**Outcome 3:** First draft of SAT design brief and evaluation criteria	DFT on pages 175–7	
4	**Outcome 3:** Pinpointing areas for research; planning, carrying out research, noting sources and commencing visualisations	SAT folio – research with IP acknowledged and source quoted DFT on pages 178–80	
5	**SAC 1** (annotate a design brief and its parts, write evaluation criteria, etc.) **Outcome 3:** Further research for SAT	**SAC 1 – Assessment task**	
6	**Outcome 3:** Final design brief; evaluation criteria	SAT folio – design brief and evaluation criteria DFT on pages 175–7	
7	**Outcome 3:** Visualisations from research	SAT folio – idea development DFT on pages 182–3	
8	**Outcome 3:** Visualisations – refining ideas; design options (as presentation drawings); feedback from end-user	SAT folio – design options, selection and justification of preferred option DFT on pages 183–5	
9	**Outcome 2:** Product development in industry **Outcome 3:** Creating working drawings for product specifications, construction diagrams, patterns, templates, etc.	Activities on pages 127–30 SAT folio – working drawings DFT on page 186	
10	**Outcome 2:** Product development in industry **Outcome 3:** Planning and safety	Activities on pages 134–8 SAT folio – production steps, timeline, risk assessment DFT on pages 188–91, 193	
11	**Outcome 2:** Product development in industry **Outcome 3:** Planning	Activities on pages 139–42 SAT folio – materials list and quality measures DFT on pages 187, 192	
12	**Outcome 3:** Process trials, production, log/photos	DFT on pages 195–7	
13	**SAC 2** (a written report, response to questions or an oral presentation)	**SAC 2 – Assessment task**	
14	**Outcome 3:** Processes suitable for low or mass production	SAT folio – list of processes to manufacture your design DFT on page 194	
15–16	**Outcome 3:** Production, log/photos, completing any outstanding design folio work	Submit SAT folio for teacher review	

Once your teacher has checked this, it is important for you to keep and refer to it often.

CHAPTER 18

Designer's role, the process, factors and market research

Unit 3, Area of Study 1
- 18.1 Role of the designer .... p. 114
- 18.2 The product design process .... p. 115
- 18.3 The product design factors .... p. 117
- 18.4 Market research .... p. 117
- 18.5 Design terminology .... p. 118
- 18.6 Identifying information in a design brief .... p. 121
- 18.7 Four-part evaluation criteria .... p. 122
- 18.8 Practice runs (writing design briefs and evaluation criteria) ... p. 123
- 18.9 Research and design activities from a design brief .... p. 124

(For detailed and useful information to complete Unit 3, read through Chapters 12 and 14 of your Nelson textbook.)

18.1 ROLE OF THE DESIGNER

Looking at the work of existing **designers** can help us understand their role in designing products. Some well-known designers with websites are: Karim Rashid (New York) plastics, architecture; Marc Newson (Australia and International) product designer; Steven Blaess (product designer); and fashion designers Collette Dinnigan, Martin Grant, Karl Lagerfeld (Chanel) and Akira Isogawa.

(Suitable case studies in your Nelson textbook for resistant materials are Marcos Davidson – jeweller (pages 209–13); Jacki Staude – metalworker (pages 206–9); Cantilever (pages 135–8); Schamburg + Alvisse (pages 264–7); Marc Pascal – industrial designer (pages 367–71); Jim Hannon-Tan (pages 376–9); Büro North – product and wayfinding design (pages 131–3); Bryan Cush, Sawdust Bureau (pages 203–6). For fashion: Bodypeace (pages 238–40); FOOL (pages 243–4); and Valentino (pages 285–6). You could also read 'How designers work' on page pages 5–7.)

Activity 18.1: Designer and the end-user

A case study or class discussion is one method of examining the relationship between designers and end-users. Choose a designer, and research and answer the following questions. Share your responses with the class.

1 Name a designer and the products they design.

_____

2 Where does their work come from? (For example, commissions, market research or their own ideas about what will sell.)

_____

3 Who does this designer design for (what type of people)? Describe their end-users or **target markets**.

_____

4 Does this designer actually construct the finished products? If not, explain where their role ends.

_____

18.2 THE PRODUCT DESIGN PROCESS

Activity 18.2a: The product design process

Review any activities you completed from the 'Design fundamentals' section regarding the product design process (activities in section 1.1) and read pages 8–9 (and flick through the rest of Chapter 2) of your Nelson textbook or have your teacher explain it to you.

The four stages and the steps of the product design process are listed below in a mixed-up order. Use the template below to list the stages and place the steps in their correct stage and the order in which they are presented on page 10 of the *VCE Product Design and Technology Study Design 2018–2022*. To assist you, the four stages are written in UPPER CASE letters. Note: there is one step that consists of two dot points.

• Working drawings • EVALUATION • Production • Product evaluation • PLANNING AND PRODUCTION • Research	• INVESTIGATING AND DEFINING • Selection and justification of a preferred option • Identify user, need, problem or opportunity • DESIGN AND DEVELOPMENT (Conceptualisation)	• Visualisations – concept sketches and/or 3D models to explore ideas • Design options (presentation drawings) • Design brief • Production plan • Evaluation criteria

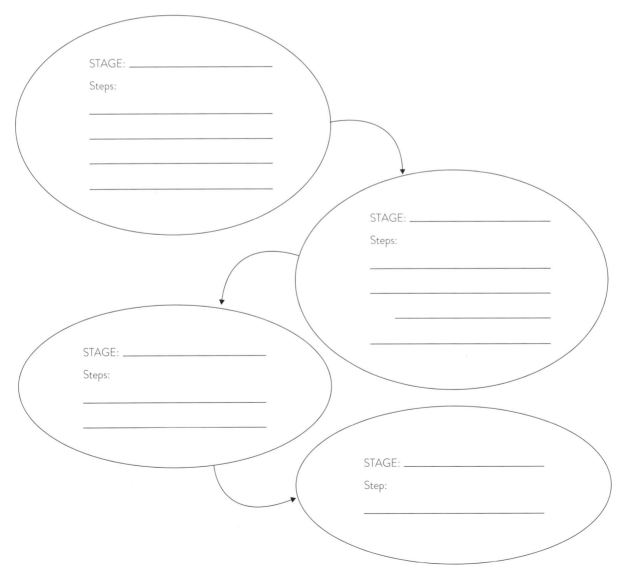

Activity 18.2b: Extension activity – reviewing design fundamentals

1. Draw your own pie diagram for the process and fill in the stages and steps.
2. Different drawing types are suitable for different purposes. They are:
   - visualisations – quick sketches of ideas for parts or the whole of the product
   - presentation drawings – 3D, rendered to show the whole product and details
   - working drawings – mostly 2D, drawn to accurate scale, dimensioned and indicating materials and construction methods.

   Refer to the table in 'Design fundamentals' section 3.1 that outlines suitable drawing types for each step on page 24. In the spaces below, draw some quick representational drawings of a simple product such as a bag, a pair of sunglasses or a child's drink cup to help you remember the drawing types and their purpose.

Draw four small **visualisations** (use black pen and colour if possible).	Develop one idea into a design option using the **presentation drawing** style (3D with more details, colour and annotated).
Create a simple **working drawing** with one view only of the product.	What do working drawings need to show? What are product specifications?

3. What components make up a production plan?

_____

_____

4. Research and **evaluation** are listed as steps 4 and 10, respectively, in the product design process, yet they are two steps that often occur throughout. Identify where else in the process (between which other steps) they might occur, and why.

_____

_____

18.3 THE PRODUCT DESIGN FACTORS

The product design factors are the factors that influence design. If you are unfamiliar with them, read Chapter 4 of your Nelson textbook and complete Activity 1.2a from the 'Design fundamentals' section of this book.

Activity 18.3: Learning the product design factors

Refer to the table of product design factors in your Nelson textbook or on page 11 of the *VCE Product Design and Technology Study Design 2018–2022*.
1   In the table below, describe in your own words the parameters for each factor. In the right hand column, write any words (from the parameters) that you do not understand.

Product design factors	Parameters (in my own words)	Words I don't understand
Purpose, function and context		
User-centred design		
Innovation and creativity		
Visual, tactile and aesthetic		
Sustainability		
Economics (time/cost)		
Legal responsibilities		
Materials (characteristics and properties)		
Technologies (tools, processes and manufacturing methods)		

2   Discuss the words you wrote in the last column with classmates until you are clear about their meaning in this context.
3   Copy the above table and write questions for each factor that could assist you in writing a design brief for Outcome 1, or as a basis for your end-user interview for Outcome 3.

18.4 MARKET RESEARCH

Activity 18.4: Market research methods

1   Write a brief explanation of each of the market research methods listed below and how the information collected might help a designer when designing a new product.
   - Survey

   - Interview

- Focus group of users of existing similar products

- Observation of consumer behaviour (i.e. when shopping)

- Experiments or trials of both prototypes and existing products, either first hand or with users

- Secondary research (Internet, data from sales or customers, industry or government bodies)

2 Create two survey questions for current users in relation to a specific product by the designer you chose in Activity 18.1; i.e. questions that are based on product design factors and could help them design an improved version.

3 Market research can be conducted with focus groups. Write two suggestions for how specific information from the focus group could inform the same designer (in identifying the end-user/s need in relation to a specific product).

Product design factors

4 Identify and explain how five product design factors have been addressed in one product from this designer.

18.5 DESIGN TERMINOLOGY

Before you write any design brief, it's important to get the terminology clear. Completing the previous activities on the product design process and the product design factors will assist you with this. It's also important to be clear about who you are designing for, i.e. one end-user (a particular person) or a (typical) end-user from a target market.

Activity 18.5a: A quick quiz

What is a design brief, what information does it contain and what is its role?

What are **criteria**?

What does 'to evaluate' mean?

Who is an end-user?

What is a target market?

What are constraints and considerations?

Focus: Purpose, function and context

Primary and **secondary functions** are among the parameters of this factor. (For further information on primary and secondary functions read pages 92–3 of your Nelson textbook.)

Activity 18.5b: Primary and secondary functions (suitable for all units)

Identify three secondary functions of the two products shown below and on the next page. The primary function has already been given and some parts of the product are listed in the first column. It is helpful to have a verb and a noun to describe the function in the second column. Add other secondary functions that you can think of.

Product	Lemon squeezer
Primary function	To squeeze juice from a lemon
Primary parts	Secondary functions
Furrows	
Spout	
Base	

Product	Straw broom
Primary function	To sweep surfaces
Primary parts	**Secondary functions**
Handle	
Bristles	
Stitching	

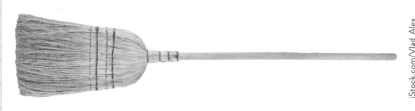

iStock.com/Vlad_Alex

Activity 18.5c: Defining the end-user

1. Think of a product similar to one that you are considering for your SAT and choose two very different end-users; e.g. an elderly person and a child of the opposite gender, e.g. an elderly man and a female child or an elderly woman and a male child.
2. Complete the following table, with ideas on what might suit these very different end-users.

Product: _____

Primary function: _____

	1:	2:
End-users		
Secondary functions that might be needed:		
Visual, tactile and aesthetic factors (name colours, textures [the feel], shapes, lines, patterns, themes, visual weight, space, forms that might appeal)		
Material (characteristics or properties to suit)		

18.6 IDENTIFYING INFORMATION IN A DESIGN BRIEF

Activity 18.6a: Annotate a design brief

Annotate the following design brief to show:
- the different sections of a design brief and the type of information each section holds
- the product design factors that are evident.

Optional: use different colours to highlight or circle the product design factors. For example, yellow for all details regarding the situation, quality and functional requirements (purpose, function and context), blue for appearance and touch (visual, tactile and aesthetic) requirements, red for material requirements, green for sustainability requirements, etc. Create a key to explain the colour coding.

Design brief

Outline of situation

Everyone in my family complains about the size of our tent when we go camping. There are five of us altogether, mum and dad, one sister, one brother and myself. We like to go camping in the mountains in spring and autumn and we like to camp where it's isolated. We share the carrying of the tent and all the other bits and pieces needed. I'm going to design a suitable tent and make a prototype to send off to a manufacturer for a production run of 500 to see how it sells. I would like to explore the idea of not having any tent poles or something innovative in that area.

Constraints and considerations

- Any material must be lightweight, strong, windproof and waterproof.
- The tent must allow for privacy for males/females at times.
- It must have easy entrance/exit access and 'windows'.
- It must keep out insects and animals.
- It needs to be bright colours.
- It needs to be very well made, all seams and fasteners reliable
- It must be sustainable in some way, either recycled/recyclable materials or long-lasting.
- A prototype must be finished before winter.
- The budget is $135.

Writing design briefs from scenarios

A scenario is a brief outline of a design **problem**, opportunity or need, from which you will write a more detailed design brief. It is important that you practise writing design briefs to assist you in creating a well-formed brief for your major project (SAT). You can write design briefs for products that are outside your material category to avoid confusion with your SAT brief and to assist in preparing you for the end-of-year exam.

Remember, a design brief is brief, not overly detailed, and does not describe the finished product – it describes the design opportunity or problem. However, it must have enough information to help the designer to create a great solution. You won't be given many details in a scenario, so you need to elaborate on this and think of all the information that a designer would need to include. Be sure to cover each of the relevant product design factors, including:

- functional requirements (including those related to the user's needs: size, performance, **ergonomics** and safety)
- visual, tactile or aesthetic requirements (the appearance or particular **design elements** and **design principles** that would be preferred either by the client or typical end-users)
- material requirements (specified materials to be used or specific **characteristics and properties** required of the materials) and any sustainability issues to be considered
- time – needs to be finished by …
- budget for, or source of, materials
- quality – belongs to the purpose, function and context factor but can also be part of sustainability (long-lasting) or legal requirements.

Cover other product design factors. There are several scenarios in Activity 18.8 to choose from.

18.7 FOUR-PART EVALUATION CRITERIA

Read about evaluation criteria on pages 19–21 of your Nelson textbook, and go over the work you did on evaluation criteria in this workbook in Section 2.2 of the 'Design fundamentals' and in Units 1 and 2.

Remember, evaluation criteria for the product come directly from the design brief and emphasise the important requirements needed in the end product. They are written in four parts:
- as questions used to judge the finished product
- with a reason for their importance
- with actions to achieve them during design, planning and production
- with a method for checking how successfully they have been achieved in the end product.

Example of four-part evaluation criteria

No.	Evaluation criterion	Reason/relevance	Actions to achieve (suggestions)	How to judge if achieved in finished product
1	Is the material suitable for the tent?	It needs to be strong, lightweight, windproof and waterproof.	I will research materials normally used for tents; I will find out costs, buy samples and test them for strength, weight and waterproofness. I will sketch some ideas for joining this material and trial them.	When the tent is finished, I will ask my family to feel the materials, carry the tent, try it out and ask them if they think it was a suitable choice. I will also test the tent in a real-life situation in poor weather and observe how well the tent performs (looking for any wear and tear).

The information in the right-hand column helps a designer evaluate the product when it is finished. It helps you to write your evaluation report in the SAT. To make sure it is suitable for judging the completed product, start the sentence with 'When the … is finished, I will …'.

Activity 18.7: Evaluation criteria for a tent

Use the table on page 123 to write three more evaluation criteria, other than the example above, from the design brief for the tent (Activity 18.6 on the previous page) incorporating relevant product design factors. To test yourself, write criteria for other than the obvious (and easier) time and cost parameters.

Evaluation criteria for the tent

No.	Evaluation criterion	Reason/relevance	Actions to achieve	How to check finished product
1				
2				
3				

18.8 PRACTICE RUNS

Activity 18.8a: Practise writing design briefs from scenarios

Design brief scenarios

Scenario 1: VCE students at your school have been allocated a room for private study. A survey showed that 80 per cent of students would like more comfortable seating for when they are studying (reading or listening). The Product Design and Technology class has offered to present design options for them to choose from.

Scenario 2: VCE students are going on a camp for a particular activity that was identified as the most popular. They need safety gear and products (clothes, uniform or equipment) that identify them as the one group. The Product Design and Technology class has offered to present design options for specific products (identify and select one product) for the group to choose from.

Scenario 3: Create a design brief for seating (or another product) for a specific end-user. You choose whether it is for an office worker, a committee, a playground, a play, children, an elderly person, a disabled person, etc. Choose an end-user, and choose a situation that is familiar to you or that is close to a real situation.

Scenario 4: Your school has banned the use of mobile phones while students are in class, but has allowed students to carry their mobile phones with them. To assist them in enforcing this rule, teachers need a storage item in which mobile phones are placed at the start of each class and which enables the phones to be given back to their owners quickly when the class ends. The teachers would like the design to be sustainable (you need to state the manner in which it is sustainable).

1. Practise writing design briefs and evaluation criteria for the scenarios listed, using the suggested structure that follows and a new document for each scenario. Use your imagination (and knowledge) to create details and include relevant product design factors (Activity 18.3).

Practice design brief

Scenario chosen: _____

_____

_____

Outline of the context (who, why, what, where, when, etc.):

_____
_____
_____
_____

Constraints and considerations (at least five):

Factor	Constraints and considerations

2 Write the full name of the factor on the left. Highlight the factors and colour-code them (or circle and connect with arrows) with the relevant design brief information as practice for your SAC task and SAT design brief.

Activity 18.8b: Practise writing evaluation criteria

Create a four-part evaluation criteria table (as below, with a new row for each criterion) for each design brief you have written. Refer to the examples in your Nelson textbook. You can also highlight the product design factors again to enhance your understanding of them.

Evaluation criterion	Reason	Way to achieve this	Checking method

18.9 RESEARCH AND DESIGN ACTIVITIES FROM A DESIGN BRIEF

Once the product design factors have been highlighted in a design brief, you can use them to direct:
- information that will need to be researched
- how the end-user/s could be consulted (interviews or surveys)
- where inspiration for ideas might be found
- ideas to be explored using design elements and principles (in drawings, models or CAD)
- measurements required
- possible tests and trials needed to help make decisions on functional aspects or the most suitable materials and processes

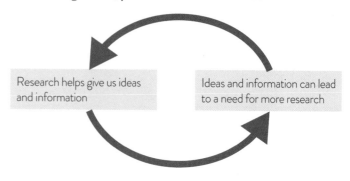

- skills needed and ways to improve them
- any other activities to assist in creating the most suitable solution.

The activities can be presented in a table, mind map or diagram (with images if applicable), or annotated on the design brief (with the relevant product design factor). Use phrases rather than single words.

Example 1: Using a graphic organiser

Copy the product design factors you highlighted in your practice design briefs or in Activity 18.3 (page 117) and add them to a table or a graphic organiser (mind map, lotus diagram, fishbone diagram, etc.). Put the name of the product (or design problem) in the middle.

Use the blank space on the next page or the mind map template on page 200 and attach/add to the relevant design brief.

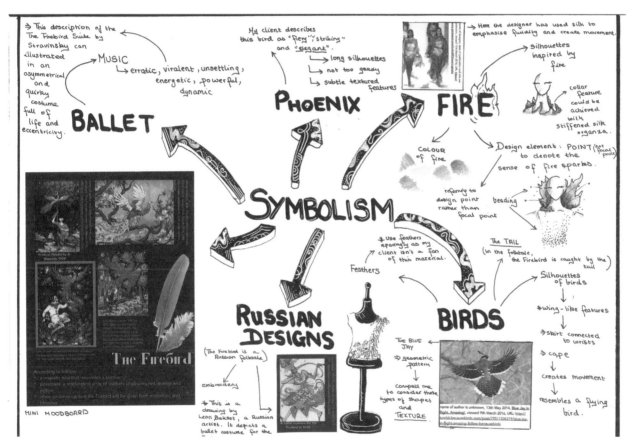

Mind map with images and drawings by student Lian Wilson to explore ideas for the visual, tactile and aesthetic factor in a costume

Example 2: Making a list or table

Copy the highlighted words or phrases into a table, similar to the one following, with a short sentence explaining what methods could be used to give the information or ideas needed to complete the product design process. Add this to your design brief document.

Word or phrase from design brief for a torch	Activities that could be carried out
Must use bright colours, material must be flexible	**Research** – get samples of the bright colours that are available in the materials and plastic to be used **Tests** – test materials to see which is flexible and suitable
The switch must be easy to use	**Research** – observe several different people use a variety of torches with different switches and make notes, take photos of switches, measure the length of the thumbs of six people and determine an average **Design** – draw 30 different ideas on the shape and placement of the switch **Trials** – borrow products with different switches and try them out

Activity 18.9: Outline research and design activities

Highlight the product design factors for the design brief for the tent (or another brief). Use these words and phrases to direct activities for the designer to help them arrive at the best solution. Use the examples and suggestions given to help you. Draw up a table as above, create a graphic organiser in the space below or use the mind map template on page 200.

CHAPTER 19

R&D, technologies, sustainability, planned obsolescence and manufacturing

Unit 3, Area of Study 2

19.1	Research and development	p. 127
19.2	New technologies, materials and processes	p. 128
19.3	The product development process and marketing	p. 130
19.4	Frameworks for sustainability	p. 134
19.5	Planned obsolescence	p. 138
19.6	Manufacturing	p. 139
19.7	Industry report	p. 140
19.8	Regulations and standards	p. 142

Read about manufacturing, sustainability and planned obsolescence on pages 348–62 of your Nelson textbook.

For this outcome, there are few 'right or wrong' answers as there are many different applications of methods and systems that apply to design and manufacturing companies.

It is useful to investigate a number of companies so that you can apply all of the knowledge in Outcome 2 to different commercial/industrial settings, i.e. specific products and companies.

Some companies that have helpful websites are listed in the table below.

Company	Where to look
Jayco – caravans	'About us' or 'Factory tour'
Taylor guitars (USA)	'Construction' (CNC Machines, Lasers, Robotics); scroll down for the movies
Yakka – workwear and footwear	Search for 'Protect'
Caroma – bathroom accessories	Go to 'Our story', then 'Research & development', then 'Sustainability'
Rip Curl – surf gear, wetsuits, accessories	Type 'RipCurl wetsuits –the lengths we go to' or 'ripcurlelearning'
Blundstone – leather boots	Search for 'Our story'
Crumpler – industrial bags	Search for 'Craftmanship'
Dyson – vacuum cleaners, washing machines	Look for 'Inside Dyson', 'Education' or 'Engineering'

Together, the following activities will guide your research and learning for Unit 3, Outcome 2.

19.1 RESEARCH AND DEVELOPMENT

(For information on research and development, see pages 348–9 of your Nelson textbook.)

Activity 19.1: The role of R&D

Research and development (R&D) occurs in manufacturing industries such as footwear and clothing, electronics, automotive, furniture and household goods, etc.

1 What is R&D? Why is it important? Refer to pages 348–9 of your Nelson textbook.

2 What sort of activities are considered R&D? Refer to any of the websites mentioned on the previous page or the case studies in your Nelson textbook.

3 How could experimenting with new technology, materials or processes help a company?

4 How is R&D different from market research?

19.2 NEW TECHNOLOGIES, MATERIALS AND PROCESSES

(For information on new and emerging technologies, see pages 350–4 of your Nelson textbook.) Additional research can be done online.

Discuss: Why is the phrase 'new and emerging' used when discussing technologies in the 21st century?

Activity 19.2: How new and emerging technologies influence design

1 Give a brief explanation of the following technologies and how they could influence product design.

Technology	Influence on product design
Laser technology	
Robotics	

Technology	Influence on product design
Computer-aided design (CAD)	
Computer-aided manufacture (CAM)	
Computer numerical control (CNC)	
Rapid 3D prototyping (3D printing)	

2. Research and explain why rapid 3D prototyping (or printing) is useful in the development stage of a product but not in its mass production.

3. Research two new materials and describe them and explain their uses. Choose examples such as wood–plastics composites, 'lightwood', carbon fibre, aluminium foam, antimicrobial copper, metal fabrics and meshes, colour-changing plastics, bioplastics, honeycomb panels, Kevlar, Gore-Tex, Nano-Tex treatments, Modal, bamboo fibre, etc. Go to the Materia website and type in the materials or category that you are interested in, or type 'new developments in fabrics' into a search engine and choose from informative websites.

4 **Extension question** (relates to SAT)
   Choose one or more of the technologies (other than 3D printing) from the previous table or a suitable industrial process that could be used to complete processes in your intended SAT product if it was to be mass-produced. Explain how the technology could be used and its implications, e.g. how it could improve your product or what might need to be changed (amount of material, complexity of the design, quality and construction techniques, etc.) to suit mass production.

_____

_____

_____

_____

19.3 THE PRODUCT DEVELOPMENT PROCESS AND MARKETING

The product development process

The product development process is about the stages a product goes through from idea (or product improvement) to 'shop shelf' when being 'mass' produced. (For information on the product development process, see pages 354–6 of your Nelson textbook.)

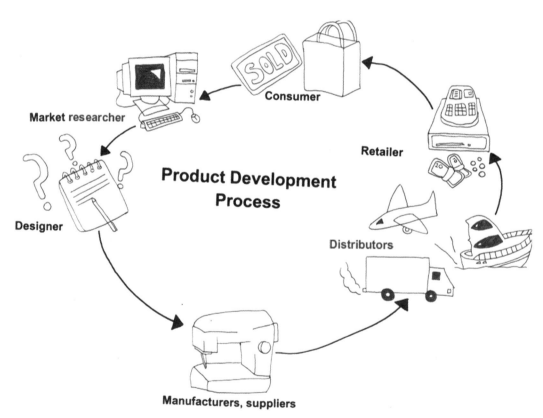

A diagram of the product development process by student Sarah Mills

Activity 19.3a: The product development process

Look at the information in the diagram on page 130 and in the diagram on page 358 of your Nelson textbook. Create your own circular diagram in the space provided by drawing cartoons for each stage in the product development process of a product similar to your SAT product, as if it was being mass-produced. Add annotations to explain stages such as quality control or outsourcing if necessary.

Label the stages with these words, placing them in the most logical order:
- design and prototyping
- retail and consumers/users
- product concept
- product evaluation and modification
- manufacturing and distribution.

Annotate the stages where market research and/or marketing would be most effective. Give some examples of methods, questions or useful information from market research that would help develop the product successfully.

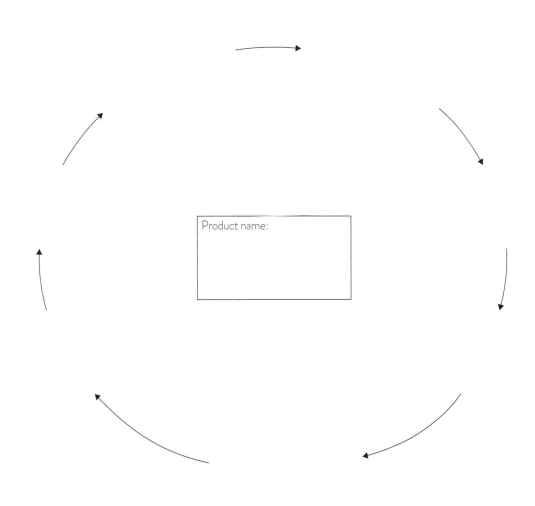

Activity 19.3b: The role of market research

(For information on market research, see page 336 and pages 357–8 of your Nelson textbook.)

1. What is market research?

___

2. How does it help companies in their product development process?

___

3. List at least four methods of obtaining information for market research.

4. Outline the differences between market research and R&D (refer to Activity 18.4: Market research methods).

___

5. Market research provides feedback (from customers, barcode sales and retailers) that is fed into the product development process. Feedback is carefully sought from retailers and customers. Technical research (on the product and technologies for manufacturing) also occurs throughout the process. How does the product development process differ from the product design process you are using to develop your one-off SAT product?

___

The 5 Ps of marketing to inform market research

Businesses target the people they want to sell to, and they make sure they have the 'right' product for the right price, sold in the right place. They also target the promotion of their product(s) to suit the needs, desires and aspirations of their likely customers and to alert them to the existence of the product. These areas are called the marketing elements or the 'marketing Ps'.

The 5 Ps of marketing are not included in Outcome 2, but they are a helpful reference for market research, which aims to define them for a particular product. The critical element of all marketing is a clear definition of the target group, which is the P for people. (For information on the 5 Ps of marketing and the product development process, see pages 357–8 of your Nelson textbook.)

Activity 19.3c: Market research elements

The following table lists the 5 Ps of marketing, which can be helpful in market research. Choose a product from one of the companies or case studies you have studied and complete the table.

1. Describe how each of these Ps is addressed in the chosen product (or, alternatively, how they could be addressed in your SAT product if produced commercially).

People (Who is this product aimed at? Who will buy it?)	
Product (What makes this product special or suitable for the target market?)	
Price (What is its retail price? Is this good value? Aimed at indicating high quality? Or priced to be seen as inexpensive?)	
Place (Where is this product sold to be easily accessible to its target market?)	List different real and virtual (online) places to sell the product.
Promotion (In what form is this product promoted to attract the attention of its target market?)	List different forms of promotion.

2 Most businesses carefully research and analyse the needs and aspirations of the 'target group' of their products. Choose a product that is made for a particular target group or 'market' (e.g. mobility scooters for the elderly) and then significantly change the target group. How would this change the following aspects?

Product _____

Current target group _____

(New) target group _____

Design of product	
Cost of the product	
Where it will be sold (place)	
Best methods of promotion	

3 What forms of market research could you use to discover useful information that would help you make the marketing decisions in each of the areas above?

___

___

4 At what stages of the product development process would this market research be useful? Explain.

___

___

5 You plan on developing your SAT product for mass production. Create three market research questions to ask potential end-users (consumers) that will give you helpful information to ensure your final prototype suits their needs.

___

___

___

19.4 FRAMEWORKS FOR SUSTAINABILITY

Refer to or complete Activities 1.2a and 1.2b in the 'Design fundamentals' section of this book.

Activity 19.4a: Sustainability

1 Designers and manufacturers can follow sustainability frameworks to guide product development. Refer to Activity 19.4b and write out the name of each of these sustainability approaches in full.

- LCA: ___
- C2C: ___
- DfD: ___
- EPR: ___

2 Distribution is defined as the way a mass-produced product is transported from factories to wholesale warehouses, retailers or online sellers. Research transport methods that are considered to be the most sustainable. Discuss and suggest methods suitable for all of these sustainability models when transporting products overseas, across land and throughout cities. Consider distance, methods, efficiency and storage. (For information on sustainability models and distribution, see Chapter 5 and pages 358–60 of your Nelson textbook.)

Activity 19.4b: Focus on sustainability frameworks

Life cycle analysis (LCA)

It is difficult to provide a detailed life cycle analysis; usually, it is done by research scientists. You have to rely on the limited information you can find, educated assumptions regarding the use of materials (e.g. a local plantation timber is better than non-plantation rainforest timber) and energy sources, and evidence of the applications of general environmental principles (e.g. reuse, reduce, recycle). Other aspects to consider include waste created during use, such as microplastics from washing synthetic clothing.

1 State briefly what is being analysed or assessed in the life cycle analysis (LCA) of a product.

_____

_____

_____

_____

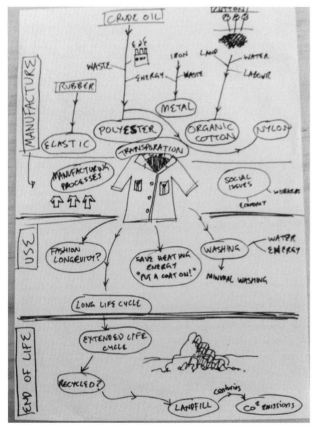

A broad LCA of a winter jacket

2 Draw a diagram in the space provided above of a broad LCA of a product from a case study in the textbook or your SAT product.

This will require research about the product's material and its correct name, which is best found from the place of purchase (or on the label of an existing product). Be sure to include any environmental impacts that come from using the product, e.g. energy use, emissions, use of water or waste by-products such as microplastics. (Refer to the environmental issues for each material category in Chapters 7 and 8 of your Nelson textbook and check the embodied energy chart on page 127.)

The example above is about a lined winter coat made with both polyester and organic cotton.

3 (Optional) Research a product that claims to have a certified LCA. Fill out the table below by rating the environmental impact with a high, medium or low in the indicated columns.
One good resource is the ebook on Ollie's Island website by Sustain Ability International Pty Ltd.

Name of product	
Material(s) being used	
Raw state of this material and country it is sourced from	
Brief description of processing and transportation involved	

		Impact (H, M or L)
Energy required to source and process materials – high, medium or low		
Type of waste produced in processing		
Energy and water required or waste created by manufacturing this type of product		
Energy and water required or waste created by the use of this type of product		
How long will the product last? Easily repaired or maintained?		
How can this product be recycled, reused or disposed of?		

In Questions 4, 6, 8 and 10, use the third column to write a marketing jingle or slogan (a phrase that emphasises the product's sustainability credentials) to be used in its promotion.

4 Explain how LCA influences design and manufacturing:

Design	Manufacturing	Marketing slogan that could be used
_____	_____	_____
_____	_____	_____

'Cradle to Cradle' (C2C)

5 What does 'Cradle to Cradle' mean? How has it developed from the 'cradle to the grave' concept? What is the main philosophy of C2C?

_____

_____

6 Explain how 'Cradle to Cradle' influences design and manufacturing:

Design	Manufacturing	Marketing slogan that could be used

Design for disassembly

7 Explain DfD.

_____

_____

8 Explain how DfD influences design and manufacturing:

Design	Manufacturing	Marketing slogan that could be used

Extended producer responsibility (EPR) or product stewardship

9 Explain EPR.

_____

_____

_____

10 Explain how EPR influences design and manufacturing:

Design	Manufacturing	Marketing slogan that could be used

11 How can consumers be involved in recycling? How can manufacturers be involved in recycling and in making it easier for consumers to do?

_____

_____

12 Thinking of a material in your category (textiles, wood, metal or plastics) and using your knowledge of sustainability frameworks, suggest some practical strategies that could be used to reduce the negative impacts of a product during the stages of its life (materials sourcing, design, manufacturing, transportation, use and disposal). (If you are having trouble, refer to pages 124–6 of your Nelson textbook.)

_____

_____

_____

13 Discuss these commonly used terms in production: 'supply chain', 'closed loop', 'circular economy'. What do they each mean?

19.5 PLANNED OBSOLESCENCE

(For information on obsolescence, see pages 363–5 of your Nelson textbook.)

Activity 19.5: Obsolescence

1 What does 'obsolete' mean? Use a dictionary if necessary.

_____

2 Why do products become obsolete? Give reasons.

_____

_____

3 Find or draw an image of a product that might be specifically affected by each kind of obsolescence and explain why it will become obsolete.

Style	Functional	Technical

4 What are the benefits and the problems of planned obsolescence for producers?

Benefits	Problems

5 What are the benefits and the problems of planned obsolescence for consumers/users (society)?

Benefits	Problems

6 What are the main negative environmental impacts created by planned obsolescence?

_____

_____

7 Discuss the differences, in terms of their sustainability, between a quality product, designed for longevity, and a product that has been planned to become obsolete within a specified time frame. Note down the main points of your discussions.

_____

_____

19.6 MANUFACTURING

Activity 19.6: Manufacturing systems and scales

1 Lean manufacturing is a system that:

_____

_____

2 How does lean manufacturing relate to flexible and responsive manufacturing?

_____

_____

3 How does lean manufacturing benefit the manufacturer?

_____

_____

4 (For information on manufacturing scales, see pages 362–4 of your Nelson textbook.)
   Give a brief explanation and an example for each manufacturing scale.

System	Explanation	Example
One-off		
Low volume (batch)		
Mass/high volume		
Continuous production (24/7)		

19.7 INDUSTRY REPORT

Activity 19.7: Revision questions for Outcome 2

This activity will help you prepare for the assessment task for Outcome 2. Write a brief report that demonstrates your understanding of the concepts, using specific examples of the products and companies you have researched and/or visited. You might need to refer to several companies to cover all the points.

1. Why is it considered important for businesses in Australia to invest in R&D?

   _____

   _____

2. Give one (or more) specific example/s of R&D in a company from the list on page 127. Check their websites for information.

   Name of company: _____

   _____

   _____

3. Explain the product development process for a commercial product from one of the companies studied.

   Name of product: _____

   _____

   _____

   _____

4. Explain why design and innovation are important for a company to stay in business.

   _____

   _____

5. Refer to a specific model of the product you chose in Question 3.

   Name of company/brand: _____

   People – describe the target market.

   _____

   _____

   Product – identify the product features that appeal to the target market.

   _____

   _____

6. Create two questions for market researchers to ask existing users of this product to give the company helpful information that could improve their next version.

   _____

   _____

7　Describe two new or emerging technologies that one of these companies uses and the benefits of those technologies for product design.

_____

_____

_____

8　Describe a product from one of these companies that has planned obsolescence, i.e. that is designed to be obsolete in a short time (rather than long-lasting); state whether it will be subject to style, functional or technical obsolescence.

_____

_____

_____

_____

9　Outline any social benefits or concerns related to production of this product, i.e. how does it help consumers/users, how does it affect workers, how does it contribute to society?

_____

_____

10　Comment on the negative sustainability issues of this product's planned obsolescence and justify your comment.

_____

_____

_____

11　Circle one of these sustainability frameworks (LCA, C2C, DfD or EPR) and explain how it could be applied to your product if it went into mass production. Identify any decisions you would need to make (in its design, materials, construction methods, quality, user instructions, disposal methods, distribution methods or promotion, etc.).

_____

_____

_____

12　Choose a designer who crafts 'one-offs' and a company producing at a low volume and explain why these manufacturing 'scales' are suitable for each and the products they make.

_____

_____

_____

13 Choose and circle one type of product from this list: a chair, a gown, a tent, a clothesline, a reading light, a coat, shoes, jewellery. Explain suitable situations where this product would be made as:

- a one-off _____

- low volume _____

- high volume. _____

14 Do you think continuous manufacturing would be suitable for this type of product? Explain why or why not.

_____

15 What are the benefits of flexible and responsive manufacturing? And how is this type of manufacturing assisted by lean manufacturing?

_____

_____

19.8 REGULATIONS AND STANDARDS

If the product you are making for your SAT has a mandatory standard, then you need to comply with this standard. Some examples are bed bunks, bean bags, children's toys, nightwear, cots, and textiles care labelling. You can check this on the Product Safety Australia website. (For information on Australian and International Standards, see pages 112–13 of your Nelson textbook.)

Activity 19.8a: International and Australian Standards

Standards are not required key knowledge in Outcome 2, but belong in the product design factor of legal responsibilities.

1 Read the page 'Benefits of Standards' on the website of Standards Australia. Then write a brief explanation of the benefits of standards in your own words.

_____

_____

_____

_____

_____

2 What does 'ISO' stand for?

_____

3 Why does Standards Australia align our standards with ISO standards as much as possible?

_____

_____

4   How do standards ensure the safety, consistency and quality of products?

_____

_____

5   What are mandatory standards?

_____

6   One voluntary standard that has mandatory sections in Australia is AS/NZS 1957:1998 Textiles – care labelling. Explain briefly its purpose, its benefits and what is required.

_____

_____

_____

_____

> **TIP**
>
> In your SAT folio for Unit 4, you are required to create a care label for your product. For details, go to Product Safety Australia and use the search facility to find out more about AS/NZS 1957:1998 Textiles – care labelling (applicable to any product with textiles, leather etc.).

7   Research one Australian Standard that is relevant for product/s of a company you investigated. Write its title out in full and give a brief description.

_____

Activity 19.8b: OH&S in manufacturing

OH&S is not required key knowledge in Outcome 2, but belongs in the product design factor of legal responsibilities. Relevant websites for this activity are Worksafe Victoria and Safe Work Australia. (For information on OH&S, see pages 41 and 113 of your Nelson textbook.)

1   What do 'OH&S' and 'WHS' stand for?

_____

_____

2   Why is OH&S important? Who is it aimed at and whom does it aim to protect?

_____

_____

_____

_____

CHAPTER 20

Activities to prepare for the SAT

Unit 3, Area of Study 3

Outcome 3 of Unit 3 covers the design, planning and commencement of production of your product. This work forms the major part of your SAT design folio. All of the work required for this outcome is outlined in the Design Folio Template in this workbook (pages 169–200).

(For further help with developing important aspects of your SAT design folio, you can refer to previous activities in this workbook, and to Chapters 2, 3, 4 and 14 in your Nelson textbook.)

Relevant workbook activities for SAT design folio work from Units 1, 2 and 3	
Unit 1	1.1 The product design process 1.2 Product design factors 2.1–2.2 Developing design briefs and evaluation criteria 2.3–2.7 Research, materials, sustainability, tests and trials 3.2–3.8 Drawing types, techniques and design activities 4.2 Intellectual property – forms of IP and attributing IP of others 5.3 Creative and critical thinking
Unit 2	11.1 Product design factors 11.2 User-centred research methods 12.5 Examples of evaluation criteria 13.1 Primary and secondary sources of research 13.2 Planning for research 13.4 Materials research, testing and process trials 14.2–14.3 Creating a style, use of mood boards and designing creatively
Unit 3	18.2–18.9 The product design process, the product design factors, design brief, evaluation criteria and planning research and design activities

Design thinking strategies in your folio

Throughout your folio, you need to use a range of critical and creative thinking techniques. Here is a loose listing of the tasks that match each.

Creative thinking	Critical thinking
• Brainstorming • Use of graphic organisers • Exploration of design elements and principles • Developing design ideas (visualisations and design options) with SCAMPER or trigger words. • Incorporating end-user feedback and input • Combining ideas in unexpected ways • Suggesting new ways to use materials and processes • Coming up with something completely new and different, unlike the 'normal'	• Use of graphic organisers to make decisions • Defining the design situation through constraints and considerations • Seeking, interpreting and implementing end-user feedback and input • Developing evaluation criteria • Researching to find and check accuracy of information • Annotating design ideas to reflect your judgement on their feasibility and/or usefulness • Selecting the preferred option • Material testing and process trialing • Production plan decisions • Analysing the finished product using evaluation criteria

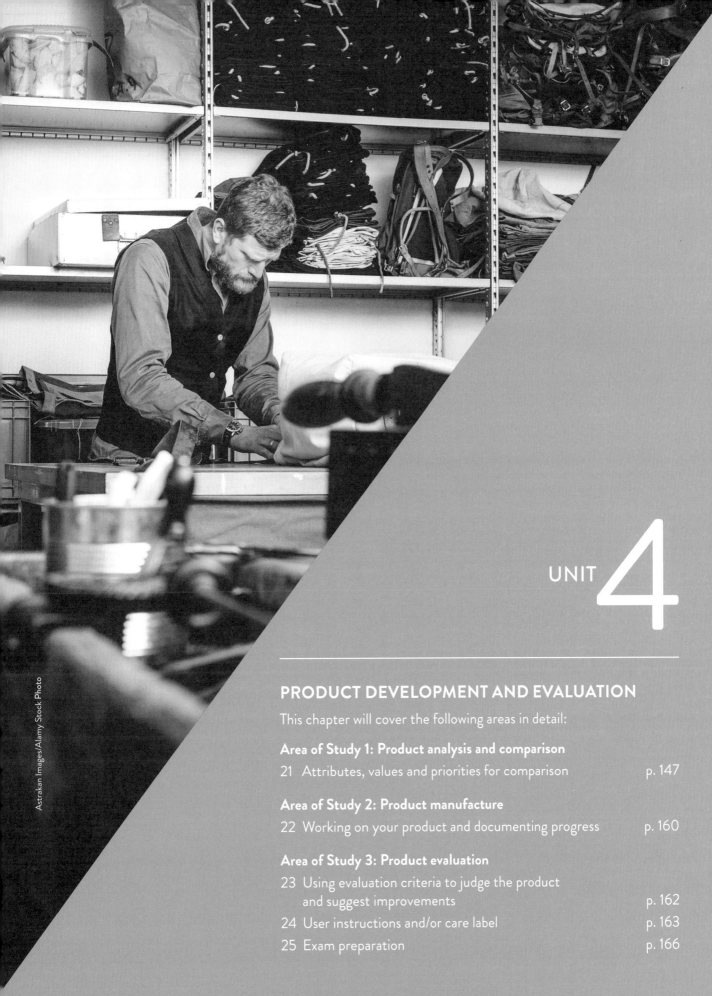

UNIT 4

PRODUCT DEVELOPMENT AND EVALUATION

This chapter will cover the following areas in detail:

Area of Study 1: Product analysis and comparison

21	Attributes, values and priorities for comparison	p. 147

Area of Study 2: Product manufacture

22	Working on your product and documenting progress	p. 160

Area of Study 3: Product evaluation

23	Using evaluation criteria to judge the product and suggest improvements	p. 162
24	User instructions and/or care label	p. 163
25	Exam preparation	p. 166

UNIT 4 CALENDAR

Week	Topic	Work for assessment	Due date
1	**Outcome 1** Product analysis and comparison activities – purpose, appeal, quality and value judgements, both qualitative and quantitative Product attributes	Complete activities on pages 148–51	
2	Environmental, social and economic issues and quality of products Valuing and priorities Practice analysis of three commercial products	Complete activities on pages 151–9	
3	Assessment – analysis and comparison of 2–3 unfamiliar products	**SAC – school-assessed task**	
1–9	**Outcome 2** Production work – focus on: • risk management • appropriate use of materials • competency with processes • accuracy with equipment • production and quality management	SAT folio: • journal with photos • log of client contact • modification sheet • finished product DFT on pages 195–7	
10	**Outcome 3** Product evaluation using evaluation criteria, checking method and end-user feedback Suggested improvements	DFT on page 198	
11	User instructions Care label	SAT folio: • evaluation report • user instructions or care label Submit complete SAT folio DFT on pages 198–9	
12	Finishing off all SAT items	**Submit completed SAT**	
13–14	Exam preparation		

CHAPTER 21
Attributes, values and priorities for comparison

Unit 4, Area of Study 1
21.1 Qualitative and quantitative information .... p. 148
21.2 Attributes of commercial products .... p. 151
21.3 Values and prioities .... p. 158

(For detailed and helpful information about Unit 4, read through Chapters 13, 14 and 15 of your Nelson textbook.)

In Unit 4, Outcome 1, when analysing a range of commercially made products, it helps to think of yourself as a consumer. Consider:

Optional activity: Product comparison – class discussion

Discuss the backpacks shown here (or use your own), or another group of products, using the above questions.

Comparing backpacks

To analyse products, you can draw on skills and knowledge that you have developed in other parts of the course, particularly around the product design factors.

You learnt about:
- both primary and secondary functions of a product
- visual, **tactile** and **aesthetic** factors
- choice of **materials** and **quality** (Unit 1, Outcome 1)
- user-centred design factors (Unit 2, Outcome 1)
- the needs of the user (Unit 3, Outcome 1) and the **context** of the product's use
- sustainability (environmental, social and economic) issues related to product design (Unit 3, Outcome 2)
- market research and how it seeks to define what consumers want
- how to use **evaluation criteria** for selecting the best design option in all units.

In this outcome, the information above will be put into practice as you evaluate the **attributes** of similar commercial products. You will also be looking at qualitative and quantitative information from research, comparative testing, emotional reaction, desire and attachment to products and how this changes.

This outcome will also give you greater skills with which to effectively evaluate your own finished product.

21.1 QUALITATIVE AND QUANTITATIVE INFORMATION

A trick to remembering the difference between these two forms of information is to think that quantitative has the word 'quantity', which refers to an amount. Therefore, quantitative refers to information that is measureable and can easily be shown in graphs or charts.

Qualitative contains the word 'quality', which is often information based on matters of opinion and is very dependent on the experience, taste, views and expectations of the person evaluating. It is also information that usually has more depth. (For information on qualitative and quantitative evaluation methods, product attributes and the values placed on them, see pages 384–8 of your Nelson textbook.)

Activity 21.1a: Commercial product evaluation methods

1 Explain user trials and how they would benefit a **designer** and consumers.

_____

_____

_____

_____

2 Give an example of comparative testing for a product (consider how the consumer advocate group Choice tests products).

Name of product: _____

Comparative testing method: _____

_____

_____

_____

3 Circle the methods below that would be considered to give qualitative information and underline those considered to give quantitative information.
- analysing sales data
- observing people as they perform a task

- asking for a score of 1–10 on how users like a product's functions
- asking users for a wishlist in a product
- creating graphs to show participants in a sporting activity
- organising a focus group to discuss a product type
- asking users to keep a diary related to use of a product
- testing a product and measuring the number of uses it can withstand
- asking questions about a product with yes or no answers.

4 Create a question for user trials to give information that is:
- qualitative

_____

_____

- quantitative.

_____

_____

5 Which form of research data is best suited for analysing large numbers of pieces of feedback? Explain.

_____

Activity 21.1b: Evaluate your favourite product

Evaluate a product that you own and love by explaining its attributes (qualities or features that are inherent in the product). Relate as many of the product design factors listed below to your product as you can, using the questions to help you.

Avoid products that have 'computer' functions such as a mobile phone, as the purpose of this task is not to evaluate digital services but the product design factors related to this subject.

1 Name of the product: _____

Purpose, function and context

2 What is the primary function of this product? What is its specific purpose? What secondary functions support the primary function to make it suitable for its specific purpose? What is special about this product that leads you to prefer it over other versions of the same product? What do you use it for? Why does it suit you so well? Where and in what types of situations do you use the product?

_____

_____

_____

User-centred design

3 Explain any sizing, ergonomic, safety or performance features or attributes that make this product suitable for your use. How 'user-friendly' is its design? What is your emotional attachment to it? How does it contribute to your wellbeing or quality of life?

_____

_____

Innovation and creativity

4 Is there anything innovative (new and different from other versions), special or clever about this product that is different from other similar products on the market? Is there something creative in this product that you like?

_____

_____

_____

Visual, tactile and aesthetic

5 Describe the visual, tactile and/or aesthetic attributes of the product that appeal to you.

_____

_____

_____

Sustainability

6 How sustainable is this product or what sustainability strategies have been used in its creation? Explain.

_____

_____

_____

Economics: Time and cost

7 Was this an expensive or cheaper product? Was it worth the money? Did you buy it when it was first on the market? Or did you wait until most people had a product of this type? Explain how it adds value to your life.

_____

_____

_____

Legal responsibilities

8 Does the product need to conform to an Australian Standard? If so, in what area? Are there any safety issues for you with this product? Do you think it might have a patent or design registration?

_____

_____

Materials – characteristics and properties

9 What is the product made from? Explain how the materials suit its purpose and use. Comment on their sustainability if not mentioned earlier.

_____

_____

_____

Technologies

10 How has the product been made? What technologies do you think have been used in its construction? Does the function of the product utilise any new technologies?

_____

_____

_____

Quality

11 How would you rate the quality of this product? What contributes to giving it this level of quality?

_____

_____

_____

Personal attachment

12 Comment on your emotional reasons for wanting this product. Will you still want it after a certain period of time? Was its newness part of your attraction to it? How will (or do) you feel when newer versions of this product come onto the market?

_____

_____

_____

13 Sum up the main reasons why you love this product. Why is it of value to you?

_____

_____

_____

21.2 ATTRIBUTES OF COMMERCIAL PRODUCTS

Product attributes

(For information on product design factors, see Chapter 4 of your Nelson textbook. For information on product attributes, see pages 381–4.)

Attributes are the features and characteristics/qualities of a *real product* (rather than a design). The attributes of a product are the tangible ways in which it addresses the **product design factors**.

Many of the product design factors have parameters that overlap and therefore attributes may not strictly belong to one factor. For example:

- The function, purpose and context of a product can overlap with **user-centred design** factors. How a product functions and what it is expected to do may require the use of **ergonomic** data – considering how the body works and interacts with the product. *Safety* may also be dependent on this information as well as the quality of the product.
- The choice of materials in a product can affect its visual, tactile and aesthetic characteristics, its function, its *quality*, and can have a big influence on the product's **sustainability**.

- A product's *quality* (which is a parameter of the purpose, function and context factor) is dependent on aspects from several factors: the way it performs its function; materials used and their suitability, strength and durability; how well a product is constructed and finished (using technologies); and whether it is a well thought out (user-centred) design.

> **TIP**
>
> When analysing and evaluating, consider the main function of a product (primary) and any supporting/associated functions (secondary) of the whole and its parts. For example, the primary function of a backpack is to hold personal objects (phone, wallet, keys, etc.) safely. Secondary functions could include aspects that determine how comfortable the handles or shoulder straps are, how accessible and easy the objects are to find, and the ease of use of zips, catches and fasteners.

iStock.com/kravcs

Other product design factors as attributes are:

Visual, tactile and aesthetic
- how it looks and feels – shape, colours, line, texture, balance, proportions, etc. (design elements and principles), its use of decorative features and its style

Economics
- its cost and value to the purchaser or user

Sustainability
- the use of renewable/non-renewable materials, energy, transportation
- the use of recycled or recyclable materials
- waste (toxicity and any related health concerns)
- designing for and ease of recycling, reuse or biodegradability
- quality of construction and how long the product will last (avoiding the need for replacement or more landfill)
- the social effects of a product's manufacture, both on the workers involved in its production and through its use and disposal on people (on their health, families, lifestyle, culture, natural environment etc.)
- the economic effects (benefits or profits and who loses out financially). Consider wages (are they fair?), where profits and wages go, whether the product helps the consumer make money or assists workers and if the broader community benefits financially.

> **TIP**
>
> You might not know all the information needed to help you make judgements about a product's level of sustainability. You should combine your research with some educated guesses. Look for authorised sustainability certification of products – examples include: FSC, Fairtrade, Ethical Clothing Australia, Good Environmental Choice Australia (GECA) and Better Cotton Initiative (BCI). Identify where the product is made and consider how workers are treated there (safety, pay and rights).

(For information about life cycle thinking and sustainability, see pages 124–4 and pages 104–7 of your Nelson textbook.)

Useful resources and websites include:
- *Waste Equals Food* – search online
- *The Story of Stuff* – animated documentary
- 'What is EcoDesign?' – Quick Guides on the DATTA Vic website (go to 'Resources', then 'Sustainability').

Thinking about quality

The quality of a product is often defined as whether it is 'fit for purpose' and can be determined by a number of factors, such as:
- **durability** – which is affected by:
  - the materials it is made from and their suitability
  - the choice of joins used in construction, and their precision and accuracy
  - the quality of components and mechanisms and how they are attached
  - how well the product is 'finished' or protected
- **level of performance** – how well the product does its job (functions or operates), whether it is reliable, easy to use, innovative, etc.
- **aesthetics** – how 'good' the product looks and feels; its style.

If a customer feels that one product is of better quality than others, and if quality is something they value highly, they are likely to choose it, even if it is more expensive. (For information about quality, see pages 94–5 of your Nelson textbook.)

> **TIP**
> Choose one of the two following activities to complete (21.2a or 21.2b).

Developing criteria

Imagine you need to choose one of the backpacks on page 147 for your own needs or for a particular person. Developing criteria before you make your decision is an effective way to assist you in making the best choice.

Use the table on the following pages to help you create criteria for selecting the most suitable backpack for yourself or another user.

You can apply this **process** to any commercial product where you are able to access three different versions, and information such as the materials they are made from and the prices. As suggested above, it is extremely helpful to consider who will use the product and in what situation it might be used (the context). For example: 'My family has decided to purchase a luxury watch for my aunt's 80th birthday' or 'My mother needs a new carry-on suitcase for travel'.

> **TIP**
> Turn the dot points under the heading 'Defining product attributes' in your Nelson textbook (page 378) into questions aimed at the product.

Prioritising attributes

Throughout the activities for Outcome 1, you need to think about how designers, manufacturers, end-users and owners **prioritise** the attributes of a product. This means identifying the features, characteristics and qualities that are the most important to different groups of people (what they value most), which will vary depending on what each group's interests in the product are. Section 21.3 considers this aspect in detail.

> **TIP**
> When asked to identify attributes of a product, choose aspects (or factors) that are definite, that you absolutely know about the product, rather than something imagined. For example, avoid saying that price is an attribute if you don't know the price, or that adjustability is an attribute if you are not sure that the product is adjustable.

Activity 21.2a: Using criteria to aid comparison of attributes

Use the following table to write criteria based on the product design factors, and then use your criteria to assess and/or rate each backpack on page 147 or three versions of a different product type.

Product type:
Who might use the product (describe)?
In what situation (context) might they use it?

Insert an image of each version of the product, naming the brand and model:		
1	2	3

Write your criteria for selecting the most suitable product for the user	Comment on how each version rates and score 1–5		
	Product 1	Product 2	Product 3
Criteria for function, purpose, context and user-centred design parameters: • _____ • _____ • _____ • _____ • _____			
Criteria for visual, tactile and aesthetic attributes: • _____ • _____ • _____			
Criteria for value/cost: • _____ • _____			
Criteria for sustainability: • _____ • _____ • _____			
Criteria for quality: Materials • _____ • _____ Construction, use of technology • _____ • _____			
Other criteria: • _____ • _____ • _____			
Summary of your comparison of products (i.e. the most suitable version for the user and why): _____ _____ _____ _____ _____			

Activity 21.2b: PMI product analysis (alternative)

Complete the PMI analysis below for each backpack or each version of the products you are comparing. Decide who you are comparing the products for — yourself or another specific end-user. Then compare your results with the table on page 157, 'PMI summary of product comparison'.

Copy this page three times and complete one for each product.

Plus
List the aspects of the existing design that are good and work well

Minus
List the aspects of the design that you think don't work well or don't look good

Drawing or photo of the product

Description: _____
Materials: _____
Where manufactured: _____
Cost: _____
Brand: _____

Interesting
What aspects of this design make you think, are unexpected or trigger your interest?

PMI summary of product comparison

Use this with the PMI analysis or as a stand-alone activity.

Product type: _____

End-user for whom products are being compared: _____

List your criteria for each area, using the questions as a guide	Make a statement here that clearly states which of the three products being compared best fulfils the criterion
Function What aspects of function are important for this end-user? _____ _____	Which product version functions best? Why? _____ _____
Visual, tactile and aesthetic What appeals visually etc. to this user? _____ _____	Which product version appeals most? Why? _____ _____
Quality What standard of quality would the user prefer in this type of product? _____ _____	Which product version is the most suitable quality for this user? Why? _____ _____
Sustainability Environmental, social and economic What would this user prefer in terms of sustainability? _____ _____	From what you can identify, which product version has the least/best impact (in any of the areas above)? _____ _____
Cost What price range would suit this user? _____	Which product version represents the best value and suits the user's budget? Explain why. _____
Summary Sum up all the relevant analysis above by stating clearly the attributes that are most important to the identified end-user, and conclude which product version is best suited to them. _____ _____ _____ _____	

21.3 VALUES AND PRIORITIES

Differing values

(For information about attributes and values and how they vary, see pages 382–4 of your Nelson textbook.)

Activity 21.3a: How values differ

In your analysis of your own favourite product or the backpacks earlier in this section, you stated the attributes that you or another end-user valued. Now consider how time can change the way you value the product. You can also consider how other people, such as the designer or the manufacturer, might value different things.

Product value changes over time

1. Think about your favourite product you identified in Activity 21.1b. As a consumer, you are more likely to value this product when it's new. Why? (Don't just say, 'Because it's new!' – think about what the product had or was like that made you like it and want it.)

   _____
   _____
   _____
   _____

2. After a number of years, how might your feelings about the product change?

   _____
   _____
   _____
   _____

3. Do manufacturers want you to keep your old product, or buy a new one? How do they get you to make that decision? How do they make you really want a new product?

   _____
   _____
   _____
   _____

4. When you see a newer version of your product with updates or improvements, how do you feel?

   _____
   _____

5. Explain why an expensive price tag might not deter you from buying a new version of a product.

   _____
   _____
   _____

6. If you were to purchase a new version of this product, what would you do with the old product?

   _____
   _____

Product attributes and values

There are a number of product attributes in the table below. These attributes might be valued differently by consumers, designers and manufacturers. For example, consumers might like a product that is value for money, but the manufacturer would value a product that gives them a large profit.

Activity 21.3b: Different values and priorities

When completing this activity, think about how each group (user, designer, manufacturer) prioritises values, i.e. how they value some attributes more than others and give them priority or place more importance on them. You can annotate to explain your choices of ticks in the table.

It is important to note that, when it comes to purchasing products, people have different priorities or aspects they think are important. Some of us value function, some prefer only cheap products, some prefer long-lasting quality products, some prefer new and prestigious products and are prepared to pay. Others are only attracted by the look of the product and many people only want to purchase sustainable products.

Companies also focus on particular attributes and prioritise them in order to attract (and keep) their target market and keep their brand status. They aim to build a reputation based on the same values as the target market by incorporating the favoured product attributes.

Indicate with a tick or cross in each column to show how you think each group might or might not value each attribute (put a dash in areas that are neither positive or negative).

Product attributes:	Consumer	Designer	Manufacturer
• durable and lasts a long time			
• light and easy to transport			
• made from materials that are cheap and easily sourced			
• made from materials that are sustainable			
• functions well			
• visually attractive and very distinctive			
• made with simple construction methods that are quick to complete			
• is good value			
• can be sold at a high price, high profit margin			
• fashionable (follows current trends)			
• very comfortable to use			
• multifunctional (can be used in more than one way)			
• innovative and creative in the way it is made, the way it functions or in its appearance			
• can be made consistently with reliable results			
• critically acclaimed (wins awards)			
• can be recycled easily			

Activity 21.3c: Product attributes

Insert an image of a product you like (or your favourite product from Activity 21.1b) into a blank document or on a sheet of paper.

Annotate to indicate the attributes that you would prioritise in this type of product (those that you think are the most important). Refer to the above list or parameters of the product design factors to help you. Ideally, you would annotate an attribute that can be seen in the image, i.e. it is definitely there and not imagined.

CHAPTER 22 | Working on your product and documenting progress

Unit 4, Outcome 2 – SAT
22.1 Managing production.. p. 160
22.2 Ensuring quality.. p. 160
22.3 Recording progress .... p. 161

In Unit 3, you started work on your product, referring to the specifications in your working drawings and following your detailed production plans.

In Unit 4, you will continue with that work, aiming to complete your product before the end of Term 3. You don't have much time!

You will have to work 'smart and fast' so that your product will be ready for assessment and so your product's critical finishing stages are not rushed.

22.1 MANAGING PRODUCTION

Here are some hints to help you manage the macro (seeing the production as a whole – big picture) and the micro (little picture – the next 10, 50 or 100 minutes):

- Keep a copy of your production plans (timeline, work plan, cutting list, risk assessment and quality planning charts) with you during production. Write on them, change them and cross off the stages you have accomplished as you go through production.
- Before, or at the start of each work session, go over your planning documents. Take 5–10 minutes to write a quick checklist of things you want to get done during that session.
- Talk with your teacher regularly to check whether you are working well on your task and keeping to your planned timeline.

22.2 ENSURING QUALITY

Use Plan–Do–Check–Act

Refer to the quality measures that you completed from the Design Folio Template on page 192.

The quality of your production work reveals much about the decisions you make during the design and production process.

Quality is affected by:
- careful selection of materials and equipment
- checking equipment and setting up correctly, not taking shortcuts
- trialling and practising
- working accurately and methodically at every stage of production
- checking regularly

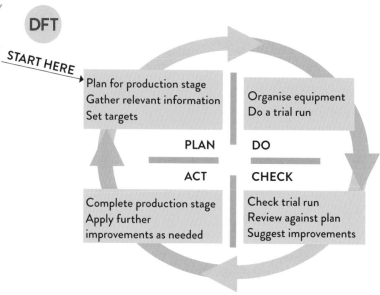

The Plan–Do–Check–Act cycle can assist with quality work

- re-thinking and re-doing processes that don't meet your expected quality standard
- adapting and changing in response to problems
- taking your time to think clearly about what you are doing – not rushing any stages and being focused. Make sure you discuss these approaches in your journal.

22.3 RECORDING PROGRESS

Photographs and explanations in your journal are critical to authenticating your production. They provide evidence:
- of your knowledge and practical experience of materials, equipment and processes
- of the quality and safety methods you used during production
- that you completed the work yourself.

Recording work in your journal

- Record your production experiences regularly while working – usually once a week. Include the date and the amount of time taken for different processes. Take photos that show you working.
- Talk about the materials, equipment, safety and quality; what went well (effective and efficient) and what could be improved.
- Plan ahead – think about the materials and equipment you need for the following week and make sure that they are available.
- Include any outsourcing or assistance received, e.g. from the teacher, a parent or an expert.
- Check your timeline – are you working to your planned dates? Do you need to plan extra work time?
- Use point form notes under appropriate headings (easier for assessors and teachers to read).
- Use a journal format that works well for you. Use or adapt the one provided on page 195 of the Design Folio Template.
- If you need to make modifications to your design during production, make sure you keep a record of any changes. Use the modification sheet on page 197 of the Design Folio Template to keep track of your changes.

Evidence of tests and trials

Continue to complete any tests or trials on materials or processes, and make their purpose clear. Make notes about what these tests, experiments or trials taught you about material characteristics or properties. Take photos of your testing and/or trialling, and add these to any research information or reports you include in your folio. Include dates and explain how this contributes to your work in creating a great solution to the design brief.

Activity 22.3: Quality construction

Careful planning, preparation, time and meticulous work will make sure that your product 'shines'. Refer to the statement of expected quality in your design brief and check that your quality measures table (page 192 of the Design Folio Template) covers the quality expected in all your processes. Practise methods needed to achieve this quality and do trials wherever possible. Add any measures or techniques that are missing.

Finishing processes are very different and those that you choose must be relevant to your type of product, how and where the product will be used and apply to all its materials.

Some finishing stages for resistant materials (wood, metal, plastics, etc.) are: checking gaps and fixing them; cleaning up any excess glues or solder, etc.; eliminating sharp or rough edges; sanding or smoothing to the required degree (mirror-like or undulating).

Some finishing stages/processes for non-resistant materials are: neatening and edging seams; top-stitching; ironing and starching; cutting all loose threads; surface treatments and some decorative techniques.

All of these processes need to be done carefully to ensure a high level of quality.

Plan the specific things you need to do to ensure an excellent finish to your product.

In your journal, document how you used quality measures to achieve the best result for all the processes involved in constructing your product.

CHAPTER 23

Using evaluation criteria to judge the product and suggest improvements

Unit 4, Outcome 3 – SAT
23.1 Evaluating your finished product..p. 162

23.1 EVALUATING YOUR FINISHED PRODUCT

Activity 23.1: Your finished product

Use the evaluation criteria developed earlier to check or test the product and to make judgements about how well it solves the initial design problem. For each criterion:
- carry out some form of testing or checking process (use the methods you identified when you first wrote your evaluation criteria), and note your results
- ask your specific end-user or possible end-users to provide feedback related to the criterion.

You can also ask end-user /s for some general comments about how well your product achieves the requirements of the brief.

Writing your report

- Write the evaluation criteria.
- Briefly state your testing/checking results in relation to each criterion and outline your user feedback – this can be done in a table (see page 196 of the Design Folio Template).
- To sum up, make a general statement about the product as a whole and comment on how well it meets the requirements of the brief. Here you can mention any other aspects of the product not covered by the evaluation criteria. Make sure that you refer to the level of quality expected in the brief.
- Identify aspects of the product that could be improved. If you identify that the quality of the product could be improved, this might lead you to reflect on activities throughout the whole process, including your research skills, your design and development, your planning and production skill, accuracy and time management.

(For information about writing your SAT evaluation report, read through pages 412–13 of your Nelson textbook.)

> **TIP**
>
> It is important to check the current SAT Assessment Criteria, published online every year in February in the *VCAA Bulletin*, to see how you will be assessed.

CHAPTER 24
User instructions and/or care label

Unit 4, Outcome 3 – SAT
24.1 Product care: User instructions and/or care label .... p. 163

24.1 PRODUCT CARE: USER INSTRUCTIONS OR CARE LABEL

(For information about user instructions and care labels, see pages 410–11 of your Nelson textbook.)

If a product's user knows how to use, care for, repair and effectively maintain their product, it will look much better and last longer. This also allows the user to foresee any future costs involved, such as dry-cleaning for textile components, or resurfacing a timber or metal product. This gives the user greater satisfaction in their product. A longer lasting product is also of benefit to the environment because it is more sustainable than a product that has a shorter life span or is affected by planned obsolescence.

Research methods of caring for, assembling, properly using and/or repairing your product. You can obtain this information:
- when researching specific details about the materials
- in the store, from information or labels provided by the supplier and on container labels of finishing substances
- from an expert (e.g. a specialised retailer, teacher)
- from other products made from the same material and your own production experience.

Read about Australian Standard 1957: 1998 at the Australian Product safety website by typing 'care labelling' into the search function. This standard applies to any product with a textile component.

Activity 24.1a: Care, maintenance, repair and/or user instructions

Describe the specific information that your end-user needs to be informed about. (There are two tables, one for products made from wood, metal and plastics, and the other for textiles products. Only fill in the areas that are relevant to your product.)

For a product made from wood, metal and/or plastics:	
Best position (or situations the product is designed for)	For example, inside, out of direct sunlight
If surface is damaged	
Best way to clean and/or polish	
Lubrication of moving parts	

For a product made from wood, metal and/or plastics:	
Chemicals to use/not to use	
Parts that can be replaced (how and where to find them)	
Other areas, e.g. when assembling or moving the product	

For a product made from fabrics:	
Fibre content	
Washing temperature	
Ironing temperature (if relevant)	
Whether the fabric can be spin/tumble dried	
Whether the fabric can be dry-cleaned	
Any special treatments required	For example, starching, special washing solution, twist drying, drying out of the sun, drying flat
Parts that can be replaced (how and where to find them)	
Other areas	

Activity 24.1b: Care label symbols (for fabric)

Care label

Remember that if you have included any fabric, it needs to be appropriately labelled. Find details of AS/NZS 1957:1998, the Australian Standard for care labelling on any textile product (including furniture upholstery).

Carry out some research to find out what each of the symbols on the next page represent. A good website is Apparel Search (enter 'care labels' into the search function).

As shown in the diagram, the five main categories of care are: wash, bleach, iron, tumble dry and dry-clean.

Make a list of the symbols and their meanings.
- List the fabrics you used in your SAT product. Research and list their laundering requirements.
- If there are different requirements for fabrics in one product, it is best to suggest the gentlest treatment in your care label (e.g. lowest temperature for washing and ironing, etc.).

Activity 24.1c: Care information

You are required to create a care label or user instructions to your end-user. There are different ways you can do this. Tick the option that would best suit your product and its use.
- [ ] A swing tag care label
- [ ] Leaflet or brochure with instructions and diagrams
- [ ] A web page with instructions and diagrams
- [ ] Video tutorials (consider how to print out if needed for an interview situation)
- [ ] Annotated image.

Use the information you have compiled from Activity 24.1a or 24.1b to create user instructions or a care label for your product.

Refer to page 199 in the Design Folio Template if necessary, and then add the care instructions to your folio.

A swing tag

Swing tag by student Steven Kearns

A care label

CHAPTER 25 | Exam preparation

Unit 4, Outcome 3 – SAT
25.1 Areas for revision .... p. 166
25.2 The exam .... p. 167
25.3 Revision activities .... p. 168

25.1 AREAS FOR REVISION

(For information on exam preparation, see Chapter 15 of your Nelson textbook.)

Questions and tasks in the exam may be based on:
- any Outcomes in Units 3 and 4
- the cross-study specifications (product design process, product design factors and materials)
- Glossary information from the Study Design.

You will need to revise, making sure you clearly understand, and can apply or use, knowledge related to the topic areas in the following tables.

The product design process

Investigating and defining	Identifying end-user need, problem or opportunity • forms of research, including market research
	The design brief • purpose, structure, parts, context, end-user, constraints and considerations
	Evaluation criteria • four parts • creating suitable questions that directly reflect design brief requirements • explaining their importance and ways they could be achieved • devising suitable checking methods to determine how well the product satisfied the criteria
	Research • planning, areas to research, connection to the product design factors, primary and secondary sources
Design and development	Visualisations to explore ideas • sketching, modelling, drawing information/inspiration from research
	Presentation drawing styles for viable solutions and to show how the whole product would look • annotation, fulfilling the needs of the brief • exploring the visual, tactile and aesthetic parameters • use of evaluation criteria to select the preferred option, and a justification
	Working drawings to provide specifications for manufacture/production • appropriate drawing styles, technical details
Planning and production	Production planning • components (timeline, work plan, materials list, risk assessment, quality measures)
	Production • knowledge of materials, tools and equipment, journal, safety • knowledge of equivalent processes used in low-volume or mass production

UNIT 4: PRODUCT DEVELOPMENT AND EVALUATION

Evaluation of the product	Product evaluation • use of evaluation criteria, methods of testing and checking, user feedback
	Care information or user instructions • relevant to materials and product type
Tests and trials	You should also be familiar with tests and trials to provide information on materials and processes. Become familiar with the words used to describe characteristics and properties of materials; what tests/trials would be suitable to check or determine these; and then what decisions this would help you to make. You need to able to state at what stage in the process you would perform these tests and trials, and how this would inform the next stage of your work.

Product design factors and their parameters

- Purpose, function and context
- User-centred design
- Innovation and creativity
- Visual, tactile and aesthetic (design elements and principles)
- Sustainability
- Economics (time and labour)
- Legal responsibilities
- Materials – characteristics and properties
- Technologies – tools, processing and manufacturing methods

Refer to the Study Design (page 11) for the parameters of the product design factors. The parameters give detailed information about what each factor covers and you are likely to be examined on them.

SAC Outcomes content

Unit 3: Outcome 1 Designing for end-user/s in product development	• Identifying needs • Use of market research • Roles of the designer and end-user/s • Use of the product design process, particularly the design brief and evaluation criteria and incorporating the product design factors • Using the design brief to direct research and design activities
Unit 3: Outcome 2 Product development in industry	• Research and development (R&D) • New and emerging technologies (laser, robotics, CAD, CAM, CNC, rapid 3D prototyping) • Lean manufacturing • The product development process and market research • Sustainability (LCA, C2C, DfD, EPR) • Planned obsolescence (style, technical, functional) • Scales of manufacturing (one-off, low-volume, mass/high-volume, continuous)
Unit 4: Outcome 1 Product analysis and comparison	• Qualitative and quantitative research • Sustainability issues (environmental, economic, social) influencing consumer choice • Valuing product attributes (differences between groups, changes over the life of a product) • Quality, its key factors and aspects

25.2 THE EXAM

- You have 15 minutes to read over the examination paper and then 90 minutes to complete your responses.
- You can bring in pens, pencils, highlighters, erasers, sharpeners, rulers, coloured pencils, water-based markers. Remember that you will be drawing designs – take in equipment that will give your designs impact.
- The exam has two sections:
  - **Section A** – short-answer and extended-answer questions, possibly multiple-choice questions, may be based on an example product (photo and brief description given)
  - **Section B** – written and drawn responses to a design brief or scenario (provided in an insert).

You will need to show your understanding of the 'knowledge' areas covered in the revision section on the previous page, and a practical use of the design process and design development skills.

Read questions carefully

The terms used in an examination question give you guidance about the sort of answer you are expected to provide, its length and depth. Make sure you read the question carefully and note the direction verbs that are used – it might help to highlight them, particularly in your practice exams.

Length and depth of response

As you read through an exam, note how much space and how many marks are allocated to each question. This can indicate roughly how long should be spent on the question. There are usually about 85–100 marks allocated for an exam in total, so that approximately translates to a minute per mark but this includes reading and interpreting the question and formulating your response.

The marks can also indicate the depth of your response. A single mark question usually requires one main point as a response, or a couple of briefly listed words. Questions that have two marks allocated often require two major points in the response, or a response that is more detailed and explanatory.

Question formats

There are a number of standard types of questions, and it is good to recognise them and know how to answer them effectively. Note that not all forms of questions are used in every exam. Question formats might include:
- multiple-choice questions
- a number of short-answer questions grouped around a concept
- 'identify and explain' questions
- extended-response questions (worth 6 marks)
- responding to a brief
- written answers
- drawn design option and process diagrams.

(For information on exam questions, see pages 420–5 of your Nelson textbook.)

Activity 25.2: Questions

Download one of the exams from the VCAA website (go to 'Students', then 'VCE studies', then 'Product Design and Technology', then 'Examination reports'). Before you attempt to do the exam as a trial, quickly read through the questions and use highlighters of different colours to identify the different types of question formats.

25.3 REVISION ACTIVITIES

Complete a number of the following activities. Concentrate on the content areas that you are **less sure of**.

Activity 25.3a: Thinking visually

Sometimes it helps you to understand an area if you can remember it visually. Develop a diagram or a visual representation that explains each of the following concepts:
- the product design process (related to a circular shape) and nine product design factors
- the structure of a design brief (including outline – constraints and considerations), evaluation criteria and their relationship to the brief
- materials categories and classification
- the product development process – used to develop products in industry
- the stages in the life cycle of a product related to environmental impact.

Activity 25.3b: Mind maps

Develop a mind map for each of the topic areas listed in the 'Areas for revision' tables (25.1, pages 166–7). Use the title of each table as the centre of the map, list the main content areas and provide further details of each area. Print and photocopy the mind map outline template provided on page 200 as many times as required for the topics covered.

Activity 25.3c: Practise drawing!

- Practise drawing products in your material category and enhancing your drawings by exploring the design elements and principles. Draw interesting shapes, add clear outlines and colour, show texture, add exploded sections to show detail/s, annotate the materials and construction processes. Think about how to be innovative, i.e. how to create a design that is unexpected, not the usual thing, and one that is not only new and different but is an improvement. Use the design brief questions from past exams and set a timer for 15 minutes.
- Draw several diagrams of processes involved in your SAT as preparation for the process diagram question. Set a timer for 2 minutes per diagram. Use the squares provided in past exams as indicators of the size required.

DESIGN FOLIO TEMPLATE

VCE Product Design and Technology Units 3 & 4

The following chapter covers the requirements for the school-assessed task (SAT):
Unit 3, Outcome 3 – Designing for others
Unit 4, Outcome 2 – Product manufacture
Unit 4, Outcome 3 – Product evaluation and care instructions

VISUAL CHECKLIST FOR SAT FOLIO CONTENTS

Product Design and Technology
SAT: Folio

Contents
_____
_____
_____
_____

Client or end-user profile
(1–2 A3 pages)

Could go here or after design brief.

Design brief
Outline of context
_____
_____

Constraints and considerations
• _____
• _____   Highlight product
• _____   design factors
• _____
• _____
• _____

Quality expected

Evaluation criteria:
For design options (optional)

Criteria questions

Evaluation criteria:
For finished product in 4 parts
• As a question
• Reason in brief
• Achieve by
• Check by

Q	Reason	How	Check

Research (2–6 pages)
Primary (done personally)
Interview/survey
Measurements
Trials/tests
Own photos of background
 information or inspiration

Secondary (desk research)
Inspiration (IP acknowledged)
Ergonomics (sources quoted)
Prices/availability
Materials
Sustainability (sources quoted)

Idea development (2–8 pages)
Brainstorming/graphic organisers
Visualisations
Mood boards, models, etc.

Visualisations mixed with research images and annotated

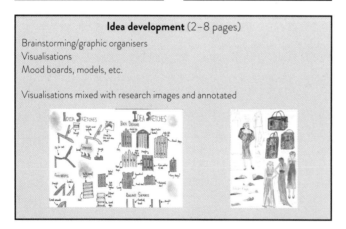

Design options (3–6), with criteria and feedback grid

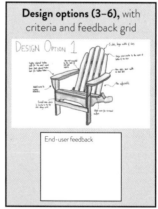

End-user feedback

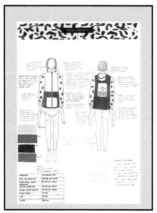

Table to select preferred option

Criteria	\multicolumn{4}{c}{Design options}

Criteria	1	2	3	4
1 Works?	5/5	4/5	3/5	5/5
2 Colour?	4/5	4/5	3/5	3/5
3 Handle?	1/5	4/5	2/5	3/5
4 Opens?	4/5	4/5	3/5	4/5
5 Theme?	4/5	4/5	5/5	5/5
Total	18/25	19/25	16/25	20/25

Justification
Clearly identified preferred option plus written **justification** paragraph, using feedback and referring to how the design drawing has met the criteria.

Several views of preferred option – this is optional but may help to calculate working drawings.

DESIGN FOLIO TEMPLATE

Working drawings 2D:
(1+ pages)
Resistant materials
Orthogonal drawing, plans, templates (may include 3D)

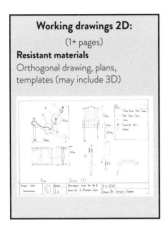

Working drawings 2D: Non-resistant materials

Garment 'flats'

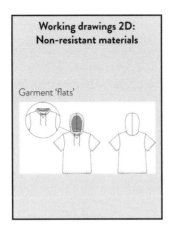

Production planning
Planned sequence (1+ pages)

	Process	Equip	Safety	Hrs
1				
2				
3				
4				
5				
6				
7				
8				
9				
10				
11				
12				
	Total time			

Production planning
Timeline (1 page)

Dates written here

(Gantt chart rows 1–12)

Production planning
Materials list and costs (½–1 page)

- _____
- _____
- _____

Quality measures (1 page)
I will achieve quality at these stages by:

Process/skill	Quality expected and/or methods
Marking material	
Cutting	
Halving joint	

Production planning
Risk assessment (1–2 pages)

Use a table to cover the main processes in your production that have a hazardous element; explain what the hazard is, the possible injury type, the risk and the controls.

Materials information, trials and tests (1–4 pages)
Could go here, in research section or after design options

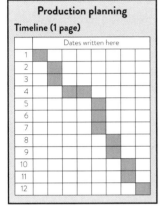

Safe operating procedures
How to set up and use major pieces of equipment and why selected

Using a router
Check the correct bit is secure and depth is accurate, secure work with clamps, pull into the cut.

Industrial processes
(To enable low volume or mass production of your product)

My processes	Industrial equivalent
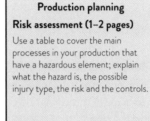	

Record of progress
Photos, log or journal (2–6 pages)

_____
_____
_____

_____
_____
_____

Modifications and feedback

Planning for next session

Evaluation report
Using evaluation criteria, explanation of checking method and reference to end-user feedback.

_____
_____
_____
_____
_____
_____

Suggested improvements
_____
_____
_____

Care label or user instructions

___ Assemble right way up, always leave tray at 90 degrees

Pure silk
Hand wash only

Blackwood and oak
Keep out of sunlight
Avoid contact with moisture

Images clockwise from top left: Steven Kearns, Nicole Crozier, Shutterstock.com/OlegSam, Shutterstock.com/OlegSam, Lian Wilson.

USING THE DESIGN FOLIO TEMPLATE

You can use these pages as a draft or rough copy and use your teacher's feedback to create final copies to be presented in a separate folio. Alternatively, you can complete work (or drafts) on separate sheets and, after your teacher's feedback, present your final copies on this template. Written aspects (e.g. journal entries) can be typed and added when complete. Some sections will need extra pages, such as those for idea development and design options. Pages may be enlarged to A3 size (141%) to suit your individual requirements (these are marked).

SAT FOLIO CHECKLIST

Investigating and defining (Unit 3, Outcome 3)

- [ ] End-user profile based on an interview
- [ ] Design brief (including quality indication)
    - Outline of the context
    - Constraints and considerations
- [ ] Evaluation criteria
    - Questions for the design options (optional)
    - Four parts for the finished product (question, justification, ways to achieve it, testing/checking of completed product)
- [ ] Graphic organisers for planning research and design activities (optional)
- [ ] Research (include both primary and secondary sources):
    - For information and inspiration
    - Sources of information indicated and IP attributed

Design developmental work (Unit 3, Outcome 3)

- [ ] Graphic organisers for exploring design ideas and directions (optional)
- [ ] Visualisations – idea and concept sketches, annotated to show influence of research; scale models
- [ ] Design options – three to six presentation drawings (3D) of different, fully worked out design concepts (annotated to show proposed materials, processes and how the option satisfies the design brief – can be scored in a grid with end-user feedback); drawn using CAD if appropriate
- [ ] Justification of the preferred option (which is clearly identified) and an explanation of how it meets the criteria, including end-user/s feedback
- [ ] Working drawings of preferred option (using CAD if appropriate)

Planning and production (Unit 3, Outcome 3 and Unit 4, Outcome 2)

- [ ] Production planning components
    - Production steps with equipment and materials needed and estimated time
    - Timeline
    - Materials cutting list and components list
    - Quality measures
    - Risk assessment
- [ ] Trialling and testing, on materials and processes being considered for the product
- [ ] Judgements on suitable tools, equipment and machines and safe operating procedures
- [ ] Journal: time and date; use of materials, equipment, processes, safety procedures and quality methods (include photos of progress); outsourcing or any assistance given; end-user feedback; modifications sheet (or documented throughout)
- [ ] List of industrial processes suitable for low volume or mass production

Evaluation (Unit 4, Outcome 3)

- [ ] Evaluation report of the product using evaluation criteria, testing/checking, end-user feedback and suggestions for improvement
- [ ] Care label or user instructions

INTERVIEW QUESTIONS

Create interview questions for the relevant product design factors to ask the end-user/s. Aim to find out what is important to them. Not all factors have to be covered.

Purpose, function context

_____

_____

User-centred design

What are the specific needs of this end-user?

_____

Innovation and creativity

_____

Visual, tactile and aesthetic

What sort of stylistic and **aesthetic** features typically appeal to this end-user? Describe them below and then collect visual examples to add to their profile.

Colours	
Lines, shapes, forms	
Patterns; textures	
Balance/proportion Symmetry/asymmetry	

Economics (time and cost)

Their budget? (Due date for finish will be set by your teacher.)

_____

Materials (characteristics and properties)

What does the material need to do?

_____

_____

Sustainability Legal responsibilities Technologies

Circle the factor and draw lines to indicate the related questions that you write below.

_____

_____

_____

_____

END-USER PROFILE

A3

Information, phrases and images relevant to your end-user, the design situation and brief. Collect a range of images of things that appeal to your end-user or the people in the user group, such as colours, textures, objects or important features of their lives. Add words that are significant. You can highlight single words, catchphrases, brand names, etc. Think of things that represent the group, or that users in the group aspire to. This can also be done to explore the interests of a single end-user. Change the title, size and shape of the boxes below to suit.

Suggestions are:

- Explanation or reason for budget limitations
- Gender, age and living situation (if relevant to the brief)
- Tastes and style (insert images of their favourite colours shapes, music, textures, objects, designers, architecture, aspects of nature, etc.)
- Interests
- Colour themes, materials, styles in the product's expected location
- Ethical, sustainability or social concerns
- Other relevant information

DESIGN BRIEF

Outline of the context (or situation)

This explains the need, opportunity, problem or design situation and the end-user. It gives the background, explains the reason and the circumstances. It should cover: **Who** is the end-user an individual user or a **target market**? (Be brief and very targeted to the design situation; this information can be expanded in the 'End-user profile'.) **Why** do they need the product? **What** do they need and what will the product be used for – what is its purpose? **Where** will the product be used (indoors/outdoors, formal/informal, specific activities, etc.)? **When** – How frequently will the product be used (special occasions, everyday use)? Use the design brief web in the 'Design fundamentals' section, on page 9, if it helps you to construct your brief.

Remember, the design brief does not describe the finished product! Start your own document if more space is needed but use the same titles and layout.

_____
_____
_____
_____
_____
_____
_____

Constraints and considerations

Design constraints and considerations are the requirements of the design situation. Highlight and colour code the relevant product design factors.

Product design factors	
• Purpose, **function** and context (functional aspects, safety, sizes, etc.) • User-centred (specific needs) • Visual, tactile and aesthetics • Materials requirements (e.g. flexible, recyclable, rigid, rustproof, etc.) • Economics (due date and budget) • Sustainability • *Innovation and creativity* • *Legal responsibilities* • *Technologies* • Expected quality **Note:** Expected standards of quality must be included in the design brief; the factors in *italics* may not be obvious in the brief.	_____ _____ _____ _____ _____ _____ _____ _____ _____ _____

9780170400404

EVALUATION CRITERIA

For the design options (6–8; optional)

You may find it helpful to create questions to help you select the preferred option. They will address the same product design factors as the questions in your four-part evaluation criteria **for the finished product** on the next page; but reword them to apply to your design option drawings. Questions could start with: 'Will this design option allow for …?', 'Can this design option be …?' and 'Does this design show …?'

	Evaluation criterion	Product design factor
1		
2		
3		
4		
5		
6		
7		
8		

Delete or add more rows as necessary.

After you have drawn the design options, you can use the questions from the above table to guide end-user feedback and your own response, using critical thinking. Some students find it helpful to add a miniature version of this grid next to each design option drawing, score the criteria and record end-user feedback in an extra column or row. See below.

Scoring each design option with a grid (optional)

	Evaluation criteria for Option 1	Score
1		/5
2		/5
3		/5
4		/5
5		/5
	Total score for Option 1	/25
	End-user feedback	

Four-part evaluation criteria for the finished product (6–8)

These are questions that will be used to evaluate the product after it is finished – they help you to judge whether the product fulfilled the needs of the end-user/s and to what degree. Aim to include reference to relevant product design factors from your design brief. Word them so that they are directed at your finished product. Example: in the last column, begin with 'When the product is finished …' or 'Is the finished product …'

Evaluation criterion (as a question)	Justification (why it is important to the design situation, etc.)	Actions to achieve in the product design process (e.g. research, measure, trial, test, practise)	How to check if achieved, and to what degree, on finished product
1			
2			
3			
4			
5			
6			
7			
8			

Add more rows as necessary.

GRAPHIC ORGANISERS

To define research and development work needed

Use as an alternative to, or in combination with, the 'Research plan' table on the next page. Choose from the appropriate orange heading to indicate your topic.

To explore ideas for the solution

Use another graphic organiser to brainstorm visual and practical ideas for your intended product. Use the mind map template on page 200 or create your own type of graphic organiser.

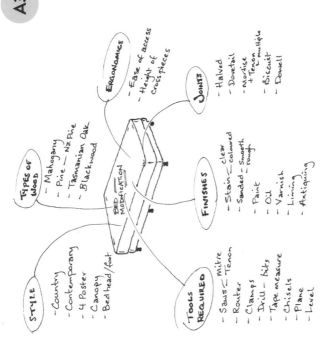

Concept map by student Jonathon Scampton

RESEARCH PLAN

Use this table in combination with your graphic organiser (defining research) from the previous page, or as an alternative to it. Research areas should come directly from your design brief requirements. Refer to the suggestions on the next page if you need help.

A3

Area for research	What I already know	What I need to check or find out	Why is this important or useful?	Where or how I will find the information?	Summary of findings	How might this affect, or be used in, your designs?

RESEARCH AND IDEA DEVELOPMENT

Create four or more A3 pages with headings for each of the following areas. Fill the pages with images, information and your own ideas – keep empty white space to less than one-third of the page.

Information, inspiration and visualisations

Complete research using both primary and secondary sources. Insert (or paste manually) images into a document and explain why they are relevant with annotations. Specific information can be added if it is relevant.

Areas that could be covered include:

- fashions, trends, styles, historical background, inspirational images from nature or other products
- issues related to the end-user's requirements or interests, or the requirements of the situation, plus important measurements
- aspects of the product's function, **ergonomics**, safety issues related to the use of the product
- construction methods, tools and equipment
- information related to materials sustainability, etc.

For secondary sources, you must make the creator's name obvious (IP and moral rights) of any 'works' and give the source of any information, diagrams or images that are not your own work or words, e.g. textbooks, website addresses or magazine titles.

For primary sources, give a clear title, explain what you did and note the date.

Leave space in your research pages to draw visualisations of your own ideas or to insert photographs of 3D models. Placing your visualisations near an inspirational image shows a clear link, which can be clarified further with annotations.

Research on materials and processes

Present evidence such as photos of your tests or trials here. These pages can be completed at this stage, after you have decided on the preferred option or during planning. If possible, attach small samples with a description of the material (including its correct name) and why you would use it. Include photos of tests performed or samples of any skills practice that you undertake throughout Units 3 and 4. Present information visually wherever possible, e.g. in graphs or diagrams. Use the materials test in Activity 2.6a, and the formats used in Activities 2.5d (Materials) and 2.6b (Processes) to structure your research in these areas. More pages can be added throughout the process.

Other research

Be sure to include other research relevant to your context, end-user/s and intended product. Give the page a clear title and annotate to explain relevance.

Visualisations: Idea development

Use one or more pages for further visualisations or design ideas. If not done previously (see instructions under the previous heading), images from your research can be inserted among your own ideas and annotated to show their influence.

Experiment with lots of creative and unusual ideas for the product as a whole, in parts or focus on details. Refer to work you have completed previously in Units 1 and 2 on the design elements and principles and designing creatively.

Incorporate patterns, unusual lines, colour and texture combinations, shapes from nature, altered proportions, adding and subtracting elements and style features as required.

Include functional aspects of the product (how it will work) and aesthetic aspects (how it will appear beautiful to the end-user).

Annotate any feedback from the end-user/s.

Set a time limit, outline your drawings boldly with a pen and use colour to bring the pages alive.

Models

You can also develop your ideas by making three-dimensional models using a range of materials (e.g. paper, cardboard, foam board, plywood, fabric). Take photos of your models and insert them into a document for your folio. Annotate them, explaining design features and things that work well (or don't work well). Include feedback from end-user/s and your own comments about the design brief requirements.

Different types of models can be created at many different stages, and can be used as either a creative or as critical thinking technique.

DESIGN FOLIO TEMPLATE

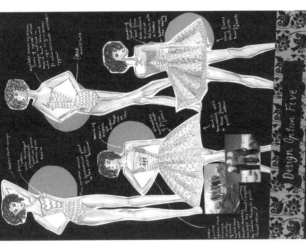

A design option drawing by student Laura McKenzie

Mood board/style guide

A mood board or style guide is a page of visual references (pictures, photos, drawings, line patterns, textures, template outlines, colour swatches etc.) and minimal text to convey a 'mood' or 'style' that would appeal visually (and aesthetically) to your end-user. It includes both creative (combining new ideas) and critical thinking (narrowing down the selection).

You can refer to it while creating your design options.

PowerPoint is a great way to create a mood board – only one slide needed! There are also some great collage apps that can be used for creating a mood board. Insert the slide into a document, give it a heading and print out on A4 or A3 paper.

DESIGN OPTIONS

These are 3D presentation drawings and are complete design ideas. Between three and six design options are required, one A3 page for each. They should include annotations to show or explain:

- how design features satisfy the evaluation criteria along with feedback from the end-user/s (see grid on page 185)
- features that have been inspired by your research
- materials and construction methods to be used
- explore the visual, tactile and aesthetic parameters, particularly colours (by using a range of media)

Optional inclusions are:

- computer drawings (must be pictorial, not 2D)
- samples or photos of materials to indicate texture, colour and pattern
- costs – materials per metre, components, etc.
- extra views, and close-up diagrams of hard-to-see or hidden details (refer to drawing section in Unit 1 of this book)
- a design option criteria scoring grid (pages 176–85).

RESEARCH AND IDEA DEVELOPMENT

Read the instructions on page 180.
This page shows my research and ideas on …

VISUALISATIONS

Read the instructions on pages 180–1 for research, visualisations and moodboards.

This page shows my own ideas for …

OPTION NUMBER: ___

Follow the instructions for design options on page 181.

Total scores for the design options (optional)

A useful (critical thinking) method to collate the end-user feedback and scores is in a grid as seen below.

Once each option is scored, transfer their totals into a larger grid such as the one below to show clearly which option achieved the highest score and to assist you in selecting the preferred option.

Add more rows/columns or create your own grid if necessary.

Evaluation criterion	Design option 1	Design option 2	Design option 3	Design option 4
1	/5	/5	/5	/5
2	/5	/5	/5	/5
3	/5	/5	/5	/5
4	/5	/5	/5	/5
5	/5	/5	/5	/5
6	/5	/5	/5	/5
Total for each option				

Justification of the preferred option

Make it clear which option is the preferred by stating it clearly and highlighting the column with the highest score. Create further drawings if required (e.g. three views or a combination of options).

Justify why the preferred option was the one selected. Explain the results of the grid and how and why this design option best satisfies the evaluation criteria and the needs of the end-user/s (with their feedback).

My preferred option is design option number _____**, which scored** _____ **out of a total of** _____.

_____

WORKING DRAWING OF THE PREFERRED OPTION

A working drawing (usually 2D) sets out all the information (product specifications) needed to construct the preferred option. It shows the details of shapes, sizes, placement, construction methods and materials, all components and hidden details. It should have enough information for another person to make the product from the drawing (and any plans that follow). It can also be computer aided.

- **Wood, metal and plastics:** An orthogonal drawing is very suitable (use any CAD program or Google Sketch-up)
- **Textiles:** A 'flat' drawing (Adobe Illustrator is useful here and there are many online tutorials to help).
If using a commercial pattern, it is expected that you make modifications (adjustments and alterations) to reflect your preferred option design.
Three-dimensional drawings can help to communicate how the product should look.

PLANNING COMPONENTS

See the SAT folio checklist on page 172 for the planning components required. Draw up or digitally create tables, or use those on the following pages, adjusting rows and columns as required, and place in your folio once completed.

MATERIALS LIST AND COSTING

The materials list gives details of what is needed to make the preferred option. It should include all of the different types of materials needed, and any components, patterns/plans, particular tools or pieces of equipment that may need to be purchased. Adjust the columns as needed to suit the information.

Wood, metal and/or plastic products

Product (brief description):							
Section or component	Amount	Material	Measurement in millimetres			Stock size and cost per metre	Total cost
			Length	Width	Thickness		

Add the space and number of rows needed

Purchased components/fittings	Amount	Where it will be used (if applicable)	Source or supplier	Cost per unit	Total cost
Total cost					$

Add the space and number of rows needed

Fabric/fibre products

Product (brief description):				
Part of the design or garment Notions, patterns, etc.	Material name (fabrics and fibres) or notion description	Amount needed	Cost per metre or per item	Total cost
Total cost				$

Add the space and number of rows needed

PRODUCTION STEPS

Explain in detail each of the steps that you predict you will follow to make your product. This should be completed before you start production work. Use the basic steps listed under the Technologies factor to help you. Add all the steps needed to complete any joins.

For the safety column, refer to and summarise specific safety controls identified in your risk assessment. (**Textiles:** If safety information is repetitive, delete the safety column and summarise the information under the table.)

Step	Description of the step or processes and materials or parts needed	Equipment, machinery, tools, etc.	Safety precautions to follow (indicate any steps that will be completed by your teacher or outsourced)	Estimated time (h)
Total estimated time				

Add further rows and pages as needed. If creating a new table, adjust columns to suit the information (i.e. narrow those columns with less information).

A3

TIMELINE FOR PRODUCTION: GANTT CHART

This table shows how the production steps will fit into the weeks available for production. In the first column, name the step, and then shade in the cells showing the weeks when that stage will be carried out. Include dates for each week and be sure to include the same number of steps as you have listed in your production steps.

Brief name of production step	Weeks (fill in the dates and shade when the step should be completed by)													
	1	2	3	4	5	6	7	8	9	10	11	12	13	14

Add further rows, columns and pages as needed to match your production steps. If creating a new table, adjust columns to suit the information (i.e. narrow the columns with less information).

RISK ASSESSMENT: PRODUCTION

A3

Analyse the materials, steps/processes and equipment you will use during production. Explain their hazards, calculate the level of risk (the chance of the hazards causing harm and a rating for seriousness and for likelihood of an incident happening) and what safety precautions you will need to reduce the risk of injury. Follow the path of the material/product from sourcing and delivery of materials, through all production and finishing stages, to delivering the completed product to the end-user. Some hazards relate to the working environment, rather than to particular equipment or processes. Include any hazardous substances that you may be using (refer to their SDS).

Step, process, material or equipment	Hazard	Possible injuries	Level of risk		Safety precautions or controls needed to minimise risk
			Likely? (H/M/L)	Serious? (H/M/L)	

Add further rows and pages as needed to cover all the production steps with hazards. If creating a new table, adjust columns to suit the information (i.e. narrow those columns with less information).

RISK ASSESSMENT: WORK ENVIRONMENT

Aspect of the work environment (e.g. lifting materials, moving equipment, storage, space, ventilation, ergonomics of work areas)	Hazard	Possible injuries	Level of risk		Safety precautions to minimise risk
			Likely? (H/M/L)	Serious? (H/M/L)	

RISK ASSESSMENT: PRODUCT

Analyse the product you have designed. Think of its different design features and explain the hazards each feature may create. Describe the design changes that could be implemented to reduce the risk of injury.

Product feature	Hazard	Possible injury	Level of risk		Design changes needed to minimise risk
			Likely? (H/M/L)	Serious? (H/M/L)	

Add further rows and pages as needed to cover all the possible product hazards. If creating a new table, adjust columns to suit the information (i.e. narrow those columns with less information).

QUALITY MEASURES

Think of all of the stages involved in making your product. List the methods needed to make your work of the highest quality.

Stage or process	Quality expected	How to achieve work of very high quality in this stage?
(combine similar steps, e.g. (1) cutting all pieces (2) using the router or overlocker)	(refer to your design brief, e.g. must be straight with no gaps)	(e.g. tips for accuracy, specific tools to use, etc.)

Add further rows as needed to cover all quality measures for your product; including those you learn as you go.

CORRECT USE OF MACHINES (SAFE OPERATING PROCEDURES)

Insert a photo of the machine or equipment into your document and annotate it to show the important functions or operating mechanisms. Refer to the machine's manual if necessary.

Make sure you consider the correct procedures for all aspects of using a machine, e.g. setting the machine up for use; the PPE you need to wear; changing or adjusting parts; holding materials when using the machine; cleaning the machine during and after use; checking the machine is in good, safe working order; packing up the machine; storage of the machine (if required).

When using this machine	What is the correct procedure? (for safety, to achieve the best result and to maintain the machine in good working order)						
Personal protective equipment (PPE) required *(tick)*							
Gloves	Face mask	Eye protection	Welding mask	Appropriate footwear	Hearing protection	Protective clothing	Face shield
Photo or drawing of machine	Potential hazards						
	Pre-operation (set-up and checks)						
	Operation (using the machine)						
	Post-operation (after using the machine)						
Competency test passed	Teacher's signature						

Create a new table/page for each of the main machines, equipment items or tools you will be using.

INDUSTRIAL PROCESSES

List the major processes to construct your preferred option and research how they would be done in an industrial setting to manufacture for mass or low volume production.

A4

Major processes in your one-off product	Relevant manufacturing processes for mass or low volume production (include images if possible)

Add further rows as needed to cover all processes that could be done industrially. Remember to refer to the relevant Assessment Criteria each year for all folio items.

JOURNAL

Date: _____ Time taken: _____

Photocopy this page or set up a Word document with a similar layout. Add journal entries (insert photos) in a new table for each session. When finished, delete unused space (or headings not addressed in a session); adjust the table to fit each session. Consider colour-coding aspects such as safety (in red), quality measures (in blue), feedback (in dark green) etc.

Detailed description of step What I did (insert photos and written explanations)	Safety procedures I followed:
	Quality measures I followed:
	Problems and what I learnt:
	Changes that I made or need to make:
	Feedback from teacher or end-user/s:
	Outsourced work or assistance I received:
	Things to do next session:

LOG OF END-USER CONTACT/FEEDBACK

Record contact you have with the end-user/s (or your teacher) to discuss the product. Contact could be made in person, by phone, fax or email. You should seek feedback at important stages during the design, production and evaluation phases of the product's development. Add further rows, as needed (or combine with your journal).

Date	Stage during the design and production process	What was discussed	Teacher or end-user feedback – decisions made

MODIFICATIONS SHEET

Whenever you make a noticeable change to the design of your preferred option (e.g. changing a material or design feature) or production plan (e.g. changing the order or a piece of equipment), you need to record them. Seek and record feedback from an end-user about modifications.

Have other procedural changes signed off by your teacher (just to check that they think the modifications are appropriate).

Wherever possible use a visual method of recording your changes, as some modifications can be explained more clearly in an annotated photo or diagram, e.g.:

- a comparison of your preferred option drawing with an annotated photo of the finished product and its changes
- a comparison of a segment of your working drawing with that segment redrawn and annotated
- a comparison of your timeline with the actual dates when steps were completed.

Date	Insert photo, diagram or altered plan of the modification	Description and/or image and reason for change	Discussed and signed off by teacher and/or end-user/s

Add rows as required or choose the most appropriate layout. This table can be drawn up on a computer with room to insert photos or diagrams (good exam practice). Adjust the columns and rows to suit the information entered.

EVALUATION OF THE PRODUCT

After you have finished making your product, you need to evaluate it. Evaluating means making judgements about how suitable the product is as a solution to the design brief and to consider improvements for future projects. Use the checking or testing methods stated within your evaluation criteria developed earlier; explain how you carried them out and how the product performed. Create this table in a document and add further rows as needed.

No.	Criterion description (copy the questions you wrote in the four-part evaluation criteria)	Describe how you checked or tested the product (e.g. I asked two end-users to carry it and get their feedback on whether it was light enough)	End-user/s feedback (include a comment/s where possible)	• Give your own judgement (make a comment) on how the product satisfied the criterion and • Make a suggestion for improvement

Add further rows as needed to cover all your evaluation criteria. This table could be submitted for assessment or you can use this information to set out your evaluation as a report. If setting out as a report, give the responses to each criterion a little heading to make your report clear to the reader.
Finish the report with a summary and a conclusion – a general statement about the product as a whole and any other suggestions for improvement.

Summary and conclusions	Suggested improvements

LOOKING AFTER YOUR PRODUCT

Care label or user instructions

Use the instructions you created in Activity 24.1a or 24.1b to create your care or user instructions. Consider creating a label that could be printed and attached to your product.

YOUR PRODUCT'S FEATURES

Annotated image

Insert an image of your finished product and annotate the features that satisfy your end-user's needs.

MIND MAP TEMPLATE

Add balloons/shapes or images as needed.